HOUSING MANAGEMENT, CONSUMERS AND CITIZENS

's College

Tenant participation has grown substantially over the last decade, following government legislation, advice from professional bodies and development agencies and promotion by all major political parties. On few housing issues is there such consensus. Yet, in practice, it is obvious that participation can mean very different things in different contexts.

This book explains why this is the case, and examines the growth of participation in the context of changes in the role of local authorities and their relationship with their electorates. These issues are examined in the first part of the book, which sets the context for exploring the roles of housing managers, councillors, tenants and tenants' associations in the second part. The book argues that the rise in arrangements for tenant participation masks considerable differences in the role played by tenants in different areas. These differences raise questions about the nature of power in the tenant–landlord relationship and more generally in the relationship between local government, citizens and consumers. These issues are examined in the final, third, part of the book.

Liz Cairncross is an independent consultant and researcher; **David Clapham** is Professor of Housing and Director of the Centre for Housing Management and Development, University of Wales, Cardiff; **Robina Goodlad** is a Senior Lecturer in Housing Studies, University of Glasgow.

HOUSING MANAGEMENT, CONSUMERS AND CITIZENS

Liz Cairncross, David Clapham and Robina Goodlad

London and New York

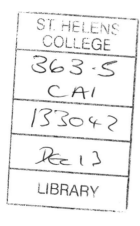
First published 1997
by Routledge
11 New Fetter Lane, London EC4P 4EE

Transferred to Digital Printing 2004

Simultaneously published in the USA and Canada
by Routledge
29 West 35th Street, New York, NY 10001

© 1997 Liz Cairncross, David Clapham & Robina Goodlad

Typeset in Adobe Garamond by
Florencetype Ltd, Stoodleigh, Devon

British Library Cataloguing in Publication Data
A catalogue record for this book is available from the British Library

Library of Congress Cataloguing in Publication Data
A catalogue record for this book has been requested

ISBN 0–415–07066–X (hbk)
ISBN 0–415–07067–8 (pbk)

CONTENTS

TABLES AND FIGURE

TABLES

FIGURE

PREFACE

Tenant participation has grown substantially over the 1980s and 1990s, following government legislation, advice from professional bodies and development agencies and promotion by all major political parties. On few housing issues is there such consensus. Yet, in practice, it is obvious that participation can mean very different things in different contexts. This book attempts to explain why that is the case, and to examine the growth of participation in the context of changes in the role of local authorities and their relationship with their electorates. These issues are examined in the first part of the book, which sets the context for exploring the roles of housing managers, councillors, tenants and tenants' associations in the second part. The book argues that the rise in arrangements for tenant participation masks considerable differences in the role played by tenants in different areas. These differences in the roles and relationships of professionals, councillors, tenants and tenants' associations raise questions about the nature of power in the tenant–landlord relationship and more generally in the relationship between local government, citizens and consumers. These issues are examined in the final, third, part of the book.

The book does not aim to be a good practice guide to tenant participation. Readers who are looking for practical advice are advised to read a book published by the Chartered Institute of Housing and the Tenant Participation Advisory Service (Cairncross, Clapham and Goodlad, 1989). This does not mean that those involved in the practice of tenant participation should be deterred from reading this book. We hope that they will gain insight into the nature of their role, and be provided with a framework which should help them to interpret their relationships with other players in the process.

ORIGINS

The book has its origins in a research project on tenant participation in council housing management which was carried out between 1987 and 1991, and was funded by the Joseph Rowntree Foundation. The research built on the results of a survey carried out by the Centre for Housing Research (Maclennan *et al.*, 1989) on behalf of the Department of the Environment. Other material, including more recent research, is drawn on and acknowledged in the text. Two main types of data emerged from our research. First, a national postal survey of all local authorities in Britain in 1986–7 elicited responses from 282 councils, representing a response of 62 per cent (which did not appear to be skewed in terms of size of authority). Second, case studies of six local authorities in Britain were selected to reflect the variety of responses which were encountered in the national survey. The case study research was carried out between 1988 and 1991.

Within the case studies, in-depth interviews were conducted with approximately 35 council officers and a similar number of active tenants' association members and councillors; a survey of tenants' associations was carried out in spring 1988, resulting in 142 responses; and a structured interview survey of 984 tenants was completed in winter 1988–9. Tenant participation meetings and other aspects of tenant–landlord relations were observed; and documentary evidence such as council minutes, newspaper articles and tenants' association newsletters was examined.

The pressure of other work has meant that it has taken longer than we envisaged to prepare the book for publication. The world does not stand still and inevitably the context within which our empirical research was undertaken has changed. However, we have been encouraged to see the book to completion by the belief that the issues raised are fundamental and enduring. The dilemmas faced by housing managers, tenants and councillors are the same, even if the way in which they are presented may alter. The precise mechanisms of policy have changed (at the time of our research the focus was on the impact of the 1988 Housing Act) but the period since 1991 has been characterised by a general continuity of policy direction, which at the time of writing seems set to survive a change of government. Therefore, we believe that the book confronts issues which are of direct relevance to housing policy and practice today.

PREFACE

ACKNOWLEDGEMENTS

We would like to thank the Rowntree Foundation who provided the grant for the research on tenant participation on which the book is based, and the Economic and Social Research Council (ESRC) for supporting the Centre for Housing Research in which the work took place. A number of people helped during the research. Jane Morton was instrumental in setting up the project and provided advice as a member of the advisory group. Janet Lewis chaired this group and other members were Ian Cole, Rory Mainwaring, Bill Hampton, Sue Robertshaw, Bruce Stafford, Gerry Smith, Marianne Hood, Mike Burbidge, Colin Ward, Peter Stringer and Robert Gormley. They gave valuable advice at all stages of the research work and Ian Cole and Bill Hampton provided extensive comments on the draft manuscript of this book. James Mitchell undertook some of the early research, particularly the postal survey of local authority arrangements for tenant participation. The ideas on which the book has been based have been presented at a number of conferences and to students at the University of Glasgow. We are grateful for the invaluable comments we received in these contexts.

Finally, we would like to thank Betty Johnstone for producing the typescript for the book. Needless to say, despite the help we received, all of the above must be absolved from responsibility for the views which are expressed here, which are our own.

Part I

THE CONTEXT

1

INTRODUCTION
The changing face of housing and local government

The subject of this book is the relationship between council tenants, housing managers and councillors. In particular, the focus is on the growing use of arrangements for tenants to participate in the management of the houses in which they live and the extent to which this empowers them. However, these arrangements cannot be viewed in isolation from the general situation of council housing. Reductions in the scale of council housing and changes in the way it is managed will, to a greater or lesser extent, influence the relationship between the parties involved in it. Councillors and housing managers who wish to retain council housing may feel more ready to listen to the demands of tenants, or they may be more likely to seek to build alliances with council tenants.

At the same time the management of council housing has been (and still is in most cases) a major activity of district and metropolitan councils. Therefore, it has to be seen in the context of general changes in local government. The last fifteen years have been a period of considerable upheaval for local government and many of the changes it has experienced have had a direct or indirect impact on council housing. Many of these changes have been the result of central government policy towards local government in recent years. It is, therefore, impossible to discuss council housing without first looking at wider changes in local government and its relationship with the centre.

The aim of this chapter is to set the scene for what follows by outlining the major trends in local government and in council housing which are likely to affect the landlord–tenant relationship. It looks first at general changes in local government, focusing particularly on relations with central government. This is followed by a review of changes in council housing which tend to be closely related

3

to and mirror more general changes. In the next section one of these changes, the growth of arrangements for tenant participation, is examined in more detail. The relationship of councillors, housing managers and tenants through tenant participation has to be viewed against the background outlined here. Finally, the chapter outlines the rest of the book.

CHANGING LOCAL GOVERNMENT

Since 1975 local government has faced attacks on its legitimacy and attempts to reshape its role which were in sharp contrast to the steady growth in its operations in the thirty years from 1945. When Crosland (then Secretary of State for the Environment) announced that 'the party was over' in 1975 he ushered in a period of instability in local government which has lasted over twenty years and shows few signs of ending.

The main changes to occur have been: consistent attempts by central government to limit local authority expenditure; the restructuring of local government through reorganisation and fragmentation into a number of different agencies often unaccountable to the local electorate; the privatisation of some local government assets and services; growing commercialism in the management of local government services; and greater involvement of users in the provision of services. Running through these changes has been an increasing central government involvement in areas (such as the school curriculum and rent-setting) which have traditionally been the province of local government.

The changes reflect the constant tension between central and local government which has characterised the period and reached its most dramatic manifestation in the anti-rate-capping campaigns of 1984 and 1985 and the abolition of the Greater London Council (GLC). To characterise this tension as a conflict purely between Conservative central government and 'radical' Labour local authorities is misleading. The relationship between local and central government is a complex one and tensions have emerged between local authorities of all political hues and central government, extending beyond disputes about spending into conflicts about the nature and management of local government itself. Here we examine the major trends under five headings, starting with public spending by local authorities.

4

Limiting local government expenditure

Attempts by central government to restrain local government expenditure have been a constant feature since the middle 1970s. The early Labour cuts were achieved by what Stoker calls 'a mix of discussion, compromise and conflict' (1991, p.12). The tone changed after 1979 when central government reduced the proportion of local authority income obtained through central grants. In 1980 a new system of central government support was introduced in England and Wales, with a block grant based on centrally determined assessments of what was needed to be spent (grant-related expenditure assessments or GRE for short). This was followed in 1981 by a system of targets and penalties to penalise local authorities who were held by central government to be overspending. When some local authorities responded by increasing the rates, central government responded by introducing 'rate-capping'. Under the 1984 Rates Act the Secretary of State for the Environment had the power to limit the rates of named authorities in England, a power which was regularly used. This was followed in 1988 by the Local Government Finance Act which introduced the Community Charge and a new system of targets and penalties. After widespread opposition to the Community Charge or 'poll tax' this in turn was abolished and the Council Tax was introduced after the 1992 general election.

The many different attempts by central government to control local government expenditure have largely been due to the failure to restrain revenue expenditures. Despite central government controls, the revenue expenditure of local authorities has increased in real terms. However, the measures have been far more successful in reducing local authorities' capital expenditure which has been cut substantially. As we shall see later, capital expenditure on housing was one of the main areas to be cut. The government's lack of success in restraining revenue expenditure highlights the fact that local authorities did not passively accept central government restrictions. Legal challenges and political campaigns were maintained although they achieved only limited success (Lansley, Goss and Wolmar, 1989). Perhaps more successful was the resort to 'creative accounting' measures. These included the build-up of revenue balances to regulate expenditure between years; the sale of assets to generate income for capital expenditure; rescheduling existing debt; capitalising expenditure such as repairs and modernisation of council property previously considered as revenue expenditure; deferred purchase schemes; and 'lease and lease back' schemes where council property was leased to another body (usually controlled by the council) in return for a lump sum usually raised from private investors. 'Creative accounting' became an uncertain game in which the government moved to close loopholes after a few innovative councils had exploited them. Nevertheless, the difficulty of controlling revenue expenditure through direct financial mechanisms caused central government to adopt other measures such as the restructuring of urban government, our second major change in local government.

Restructuring of local government

Restructuring has taken two main forms: the reorganisation of local government itself and the growing fragmentation of urban government through the creation of other bodies. Reorganisation of local government since 1979 started with the abolition of the Greater London Council and the six metropolitan county councils in 1985. Since then, local government reorganisation has been constantly on the agenda. In England a Local Government Commission was established to bring about evolutionary change by creating unitary authorities and abolishing some of the more unpopular counties created in the 1974 reorganisation in England and Wales. In Wales and Scotland, there has been a continuing concern with the structure of local government in relation to debates about devolution and independence, and in both countries unitary authorities have been created. Therefore, throughout Britain, local councils have had the spectre of reorganisation hanging over them.

Perhaps more importantly, central government has increasingly sought to bypass local authorities by the creation of, or support for, a plethora of other agencies – what Stoker (1991) has characterised as the acceleration in the growth of non-elected local government. The motivation of central government has been to reduce the role of local authorities and to channel funds through agencies which are more directly accountable at the central level and consequently less accountable to local political forces.

However, the trend towards non-elected local government has been reinforced by many local authorities themselves who have sought to involve users directly in service provision and to encourage innovation through the creation of arms-length agencies. Stoker (1991) uses a sixfold classification to cover the wide range of organisations involved. These are central government 'arms-length' agencies (such as urban development corporations, training and enterprise councils, the Housing Corporation); local authority implementation agencies (such as enterprise boards and co-operative development agencies); public–private partnerships (Housing Action Trusts, enterprise agencies); user organisations (such as housing co-operatives, opted-out schools); intergovernment forums (examples include the Glasgow Eastern Area Renewal project, and the South East Regional Planning Conference); and finally joint boards which are different from the above in that they have a legal entity in their own right (examples are the boards set up after the abolition of the GLC and the metropolitan counties in 1986).

The increasing number of such bodies has led to what can be viewed as a reversal of the Victorian trend towards the consolidation of local powers in local authorities. Rather, there has been an increasing fragmentation of powers into a large number of bodies, with different terms of reference and structures of accountability.

Privatisation of services and assets

Fragmentation has been encouraged also by the privatisation of local government assets and services, the third major trend in local government since 1979. Privatisation is a much used word but one which defies any precise definition. It is most popularly associated with the sale of local authority assets, most notably through the council tenants' Right to Buy at what can be a substantial discount. As will be outlined in the next section, this and other measures have substantially reduced the amount of housing owned and managed by local authorities. Other privatisation measures include the sale of bus companies and English Water Boards. However, apart from the sale of council houses and other assets, the commercialisation of services has been the most important way of privatising local government services, and this is considered separately.

The commercialisation of services

The most important example of commercialisation is compulsory competitive tendering (CCT). This has forced local authorities to put some of their activities (such as housing management, refuse collection or catering) out to competitive tender and to compete with private companies. The extent of competition and successful private sector bids has varied by service and area of the country. In practice, inhouse workers have won many of the contracts with only 20 to 30 per cent going to private companies. Nevertheless, in order to win the contracts inhouse workers have been forced in many cases to accept changed working practices, reductions in wages and conditions and sometimes redundancies. Therefore, the impact of CCT has been substantial in forcing on local authorities and their workers the ethos of competition, in creating incentives and a momentum towards a more commercial orientation and in fragmenting departments into 'client' and 'contractor' sections.

A commercial orientation has existed in a number of authorities for some time, but CCT has had a wide impact on authorities which previously had a more traditional public service approach. In some of these authorities, CCT has resulted in the development of a more consumer-oriented approach to service provision, providing staff with the impetus to borrow private sector language, techniques and values. At first, CCT was largely confined to blue collar local government services where the service is relatively easily defined.

However, since the early 1990s white collar services were increasingly targeted and housing management was one of the first. In 1993 pilot projects for CCT in housing management were set up in England and through a phased implementation process most local housing authorities in Britain will have been forced to go through the process by the end of the decade.

One impact of CCT has been to create within local authorities an organisational split between the purchasers (those formulating tendering briefs and monitoring the service provided) and the providers of services. In many cases the providers have been 'hived-off' to relatively autonomous 'arms-length' agencies. A clear example of the purchaser and provider split is the community care provisions of the National Health Service and Community Care Act 1990. Social services authorities act as assessors of the community care needs of individuals and families and care managers ensure that appropriate care is provided. However, the care is provided by different parts of the organisations and by private or voluntary agencies. The government has made no secret of its preference for voluntary and private agencies and incentives are built in to ensure that they are used. It is expected that a similar purchaser/provider split may well develop in housing authorities with the introduction of CCT.

Growing role of service users

The final change affecting local government to be outlined here is the growing role of consumers. Provision of local authority services is increasingly being devolved to user groups, either by local authorities themselves or through mechanisms devised by central government. An example of the latter is the provision to allow parents to vote for their school to opt out of local authority control. The running of schools is placed in the hands of a board elected by parents who have substantial powers over the day-to-day running of the school and receive their funds directly from a central government agency. For those schools which remain under local authority control the government introduced measures in the 1988 Education Reform Act to encourage open enrolment in accordance with parental preference. All larger schools in England and Wales have a system of devolved budgets and are run by local governing bodies. In addition, schools have been forced to provide performance information for parents. At the same time as parental influence has been strengthened, central government has imposed a National Curriculum in all state-funded schools and a national system of testing of all pupils at certain ages. Therefore, devolution of influence over day-to-day matters to parents and headteachers has been associated with a centralisation of funding and decisions over what is taught to central government. The consequence of these changes has been a reduction in the influence of the local education authority. Elements of this trend are present also in social services where voluntary organisations are given a greater role. And council housing provides numerous examples, as discussed later in this chapter.

SUMMARY

The 1980s and early 1990s have been a period of considerable change in British local government. Clearly not all of the changes have occurred to the same degree in all local authorities and in all service areas. Some of the changes may in certain circumstances be contradictory: obvious tensions exist, for example, between increasing the influence of users and increasing the influence of the private sector. Nevertheless, these remain the major general directions of change.

Most of these changes have been initiated by central government, although some innovative Conservative local authorities have implemented programmes which have later been taken up at a national level. At the same time there has been considerable resistance to central government plans, most notably from some Labour-controlled authorities which have sought to defend their position by mobilising support and by pursuing innovations designed to frustrate government reforms or to illustrate an alternative model of how public services can be run. Nevertheless, it is clear that since 1979 central government involvement in urban government has increased substantially in both its scope and its depth. Central government has a wider and a more detailed interest backed up by a strong battery of controls.

All of the elements discussed here are present in housing. Indeed, housing has been at the forefront of many of the changes and the battles over them. This is partly because the ideal of the 'property owning democracy' has been such a preoccupation of successive Conservative governments since 1979 and partly because control of local authority housing has been such an important element in the activities and influence of local authorities for several decades. Thus, conflict and change in local authority housing has been an example of the continuing tension between central and local government.

Housing's pre-eminent position in the changes in local government may also be partly because government activity in housing has traditionally concentrated as much on intervention in the private housing market as on the direct provision of housing. Therefore, council housing has never become a universal service in the same way as health or education. The nature of housing ensures that it is at the forefront of political debate and subject to conflicting political ideologies. On the one hand, housing can be considered suitable for production and exchange through market mechanisms. Indeed, most housing in Britain is produced and consumed in this way. However, shelter is a basic need which has severe consequences if it is not met. In addition, housing is a very expensive commodity and for most home owners, it is the single most important constituent of household wealth (Hills, 1995; Forrest and Murie, 1995). The inheritance of a house is becoming a very important mechanism for the inter-generational transmission of wealth. In addition, the benefit derived by the occupier of a house is not limited to the use of the physical structure itself, because its location determines the nature of the physical and social environment, as well as access to employment opportunities, amenities such as shops and recreational activities and public services such as schools and health care. As a consequence, housing is an important instrument for translating income inequality into disadvantage in a wider sense (Clapham, Kemp and Smith, 1990). Therefore, government intervention in housing has been considered by social democratic parties to be a key instrument in achieving greater equality in the distribution of life chances. At the same time, parties of the right have seen housing as an area ripe for the expansion of market mechanisms and a reduction in the role of government. Thus in the 1980s and early 1990s housing has been at the forefront of the changing role of local government.

THE CHANGING HOUSING SYSTEM

Central government's overall objectives for local government have been instilled into three main aims in housing policy during the 1980s and 1990s: the expansion of owner-occupation; the reduction of public expenditure on housing; and a limitation in the role of local housing authorities to one of 'enabling rather than the provision or management of housing' (Department of the Environment, 1987). The major mechanisms by which they have been achieved have been the Right to Buy, housing associations, the restructuring of council housing and the encouragement of better management performance. The growth of tenant participation needs to be seen in the light of these other changes, which are now outlined in turn.

Right to buy

All three government objectives were achieved by the Right to Buy given to council tenants in 1980. (For a review of the early years of the programme see Forrest and Murie, 1988.) Over a million and a quarter council houses have been sold to sitting tenants as a result of the programme with a corresponding decrease in local authority managed stock. At first some councils openly opposed the programme, but few continued outright opposition for long. Others chose a more covert strategy of dragging their feet over the selling process and attempted to discourage tenants by, for example, imposing such conditions on modernisation programmes that tenants would not exercise their Right to Buy. However, even this form of protest died out as central government closed loopholes and challenged recalcitrant authorities. Also, Labour local authorities had the rug pulled from under their feet by the dropping of objections to the Right to Buy policy by the national party. The Labour Party has come round to a position of support for the policy, although it would like to see councils able to make use of all capital receipts from the sales to replace the stock lost from the council sector.

Councils face restrictions on their use of receipts which have largely been used to fund capital improvements or to reduce the amount needed to be raised from borrowing or taxation at national level. Therefore, despite the long term loss incurred by the sale of assets at below market prices, and support to those buying through mortgage interest tax relief, the policy has been a support in the short term to the Exchequer. Capital receipts from the housing programme amounted to £17,580 million in the ten years after 1979. They represented 43 per cent of all the government's privatisation programmes and a key element in the public expenditure and fiscal policies of government (Forrest and Murie, 1988).

The motivation for tenants to buy their houses has been a mixture of a positive view of the benefits to be gained from home owner-ship along with a negative view of the consequences of remaining in the council sector. Clearly many tenants can take advantage of large discounts to buy their homes at well below the market value. Aided by discounts and MIRAS tax relief (now diminishing) many tenants have been able to enter owner-occupation who otherwise would have found it more difficult.

After the first surge of people buying their council homes the number reduced to a steady flow, although demand was kept high, through periodic increases in the discounts available, and the intro-duction of portable discounts (discounts can be transferred to another property other than the one the tenant currently lives in). Rent to Mortgage schemes had very little impact (Kay and Hardin, 1992). The house price boom of 1987–8 provided an additional stimulus to Right to Buy sales at a time when they might otherwise have dropped to a trickle.

Sales continue at a steady rate, but they are limited by the refusal of the government to provide a way for the more than two-thirds of tenants receiving Housing Benefit to buy, because of the public expenditure implications of such a move. The main consequence is that those able to buy are doing so and council housing is increasingly occupied by those unable to buy, consisting largely of poor and disadvantaged groups such as older people, those who are permanently sick, unemployed people and single parents on low incomes.

INTRODUCTION

Housing associations

Since 1979 the government has seen housing associations rather than local authorities as the major providers of new, subsidised, rented housing. Housing associations were perceived as a vehicle for housing provision and management which removed housing from the local political sphere. Also, they were seen by government and their tenants as being more effective and more 'user-friendly' housing managers than local authorities. Therefore, the government saw the opportunity to restrict the role of local authorities and to depoliticise the provision of rented housing, while being seen to be acting in the best interests of tenants as a whole. In fact the relative effectiveness of housing associations was largely achieved because of their favoured status as they were more highly subsidised than local authorities (Maclennan et al., 1989). The other major advantage they had was small size which is associated with tenant perceptions of effective housing management (Clapham, 1992a). However, one of the major impacts of the government reforms has been to favour large associations and encourage smaller ones to grow or to merge with others, thus negating the sector's one real competitive advantage over local authorities.

The reforms to the housing association sector were designed primarily to create a 'quasi-market' in rented housing outside the municipal sector. Tenancies were reformed along with those in the private sector to create 'assured tenancies' which offered more limited statutory security of tenure to tenants. The grant system was changed to divert risk previously borne by the Housing Corporation (or Scottish Homes or Tai Cymru) on to associations developing new schemes. (These are the central government controlled quangos which regulate and control the sectors.) Thus if building work costs more than estimated or there is a higher than estimated level of empty properties (and thus a lower than forecast level of rental income) the costs have to be borne by the association. Crucially, the level of government grant for capital costs (Housing Association Grant or HAG) is gradually being reduced over time and associations are being required to meet a growing proportion of the capital costs of new schemes from loans from private financial institutions. The consequence of this has been to force associations to increase rents or use their reserves to keep rents low, or both. In turn this has resulted in pressure from central government to keep rents low by taking the level of rents into account in the competition between

13

associations for HAG. Associations are being encouraged to diversify into the provision of housing for owner-occupation and into management of council housing after CCT. Also, they are intended to face competition from private subsidised 'licensed' landlords, although it is unclear whether this competition will materalise.

The result of these reforms has been to encourage housing to operate in a more commercial way than previously. They have to persuade private financial institutions that they are a sound investment and have to act in a more entrepreneurial fashion, accepting risk and winning against competition from other landlords. At the same time they are closely controlled by the Housing Corporation or the other central government quangos and so offer central government a more direct way of achieving its objectives than working through local authorities.

While central government has strengthened its control of the housing association movement through the Housing Corporation, Scottish Homes and Tai Cymru, downward accountability remains weak. In many associations, accountability to the local communities in which they operate and to their own tenants is limited.

Restructuring council housing

To encourage council house sales, as well as the carrot of discounts, the government has used the stick by making council housing relatively unattractive in a number of ways. Rents have increased considerably in real terms since 1979 and central government revenue subsidies to the Housing Revenue Account have been cut substantially. Also, since the 'ring fencing' of these accounts in England and Wales in 1990, subsidies from local taxation have not been possible. With an increasing number of local housing authorities making a surplus on their Housing Revenue Account without subsidy and using surplus to reduce the levels of local taxation, the government intervened to make Housing Benefit a charge on the account. In other words, Housing Benefit for low income council tenants is partly funded by the rents of other council tenants.

The impact of these measures was to put pressure on revenue expenditure on housing management and repairs which was already low in comparison with housing associations. Maclennan *et al.* (1989) and Bines *et al.* (1993) found that housing associations spend substantially more on housing management than local authorities (£278 compared with £205 per dwelling per annum in 1990–1)

(Bines *et al.*, 1993, p.29). Tenants considering whether to buy their house are faced with the likelihood of increasing rents and a deteriorating management service.

Perhaps more importantly, severe capital restrictions were imposed on local housing authorities. This has meant the virtual cessation of council house-building and the restriction of funds available to spend on modernising the existing housing stock and keeping it in good repair. Capital allocations have been increasingly targeted at projects and local authorities which follow government policy, including tenant participation in renewal such as through the Estate Action programme (Pinto, 1993).

The government have also attempted to reform local authorities by making them more market oriented and by attempting to break up the sector to remove what is viewed as their monopoly position. 'Tenant's Choice' was introduced in 1988 to allow groups of tenants to leave the sector by choosing another landlord. Opting out is a once and for all choice open only to council tenants. Housing association tenants were not given the right to opt in to the council sector.

The government appeared to believe both that existing council tenants were anxious to leave the sector and that a large number of alternative landlords, whether private companies, trusts or housing associations, were queuing up to take over the role of local councils. Neither belief proved to be well founded and there have been only a few transfers under Tenant's Choice, mostly to tenant-run trusts or co-operatives. However, the legislation had a large impact on the attitudes of housing managers and councillors as we shall see later in the book.

Increasingly common as a means of transferring council stock to landlords is the use of large scale voluntary transfer of all or part of a council's stock to existing or specially created housing associations, after a majority vote by tenants. By 1995, 40 local authorities had voluntarily transferred their stock to existing housing associations or associations set up specially for the purpose, and as many as one-fifth (21 per cent) of authorities in England had taken at least some steps towards pursuing this option by 1991 (Bines *et al.*, 1993, p.26). Conservative-controlled authorities, especially in the south of England, have been particularly active in transferring all their stock in this way. The intention is partly to take advantage of the capital grants available to housing associations and to avoid controls over local authority spending. Councillors feel they are acting in the best interests of tenants because a housing association could offer a better service. At the same time there is clearly an ideological motivation to many of the transfers, with Labour local authorities generally refusing to contemplate this option.

At first, transfers largely took the form of a transfer of all of a local authority's stock, but in 1992 central government signalled its wish that transfers should be in smaller tranches of only part of the total stock. This reflected criticism that transfers did not break up monopolies of public rented housing but merely changed who controlled them. The government appeared to want to ensure transfer to a number of smaller housing associations in the belief that this would create competition.

Another central government policy has been the creation of Housing Action Trusts (HATs). Initially aimed to be scaled down Urban Development Corporations, they were intended to take over housing from local authorities in small geographical areas and stimulate economic, social and physical regeneration. The intention was that the private sector would play a major role and it was envisaged that the houses would be taken over by private companies or housing associations. However, the government was forced to amend its plans after vociferous opposition from within Parliament and from local authorities and tenants in the areas first earmarked for HATs. As a consequence HATs have become more 'tenant friendly' with key concessions being made by the government. For example, tenants were given by the House of Lords the right to veto the creation of a HAT; were offered the opportunity of having representatives on the HAT board; and given the option of returning to the council at the end of the HAT if they so wished. As a consequence much of the opposition to HATs was withdrawn and a number have now been set up.

In Scotland there is no statutory provision for HATs and very few Tenant's Choice transfers have taken place. Voluntary transfers by local authorities have taken place in small areas of rundown council housing (between 200 and 400 properties is usual) to tenant-run co-operatives or community-based housing associations after a ballot of the tenants affected. This programme is known as Community Ownership and about 20 transfers had taken place by the middle of 1992 (Clapham and Kintrea, 1992).

Interestingly, the Community Ownership programme is supported by both a Conservative central government and local authorities which are predominantly Labour-controlled. The Right to Buy sales

16

in urban Scotland have been relatively low. Therefore, the transfer of council stock to tenant run housing associations or co-operatives through Community Ownership is perceived by central government as the most feasible way of breaking up council housing. At the same time Labour local authorities can gain access to HAG to allow modernisation of property and can give their tenants substantial control over their housing circumstances. Community Ownership is viewed by some as privatisation and by others as a form of co-operative socialism. It is therefore a programme which commands widespread acceptance across the political spectrum. Tenants have supported it largely because of the prospects it offered for capital improvements.

Improving management performance

As well as promoting the break-up of council housing the government has sought to reform the way that councils approach housing management. The government has had a continuing concern to improve what it sees as the poor housing management performance of many local housing authorities. At first this took the form of sponsoring local management initiatives on difficult-to-let estates through the Priority Estates Project (Power, 1984). Following reports by the Audit Commission (1986), by Glasgow University (Maclennan et al., 1989) and Cardiff University (Clinton et al., 1989) the government required local authorities in England and Wales to collect and publish performance indicators as a means of highlighting poor performers and focusing attention on performance. This was followed by the proposed introduction of Compulsory Competitive Tendering for housing management as a way of introducing the private sector and competition into council housing management, following its introduction in other fields of local government activity. Local authorities (not housing associations) will be forced to draw up a specification for housing management services to be provided to tenants in any area (the scale of areas chosen is limited in size in an effort to prevent local authorities trying to adopt 'anti-competitive' tactics) and then to seek tenders from their own staff and from outside agencies such as private companies or housing associations. The contract will go to the contractor offering the combination of price and quality which the authority considers best, taking account of tenants' views about standards and quality criteria (Goodlad and Rosengard, 1994). Although it seems likely that initially at least, the contracts will mostly be won by local authorities themselves, it is possible that, over a period of time, a local housing authority will have all its housing management performed by outside contractors with its role restricted to the specification of contracts and monitoring the performance of contractors.

In general, CCT has been opposed by local authorities and tenants, and represents a radical break from traditional local government methods of operation. In contrast, some measures, particularly those designed to foster decentralisation, have been promoted as much by local authorities as by central government. Two-thirds (66 per cent) of English housing associations had decentralised at least part of their housing management by 1991 (Bines *et al.*, 1993, p.25). A smaller, but still significant proportion (42 per cent), had introduced estate-based initiatives with tenant involvement (Bines *et al.*, 1993, p.26). Authorities have also embraced the new emphasis on management performance with some enthusiasm, many developing the use of performance indicators for internal use in the late 1980s before publication of a report for tenants became compulsory in 1990 (Bines *et al.* 1993, pp.146–7). The focus on tenants in these developments illustrates the important role they apparently now occupy in the view of central government and many local authorities. It is to this transformation in the tenant–landlord relationship that we now turn.

The growth of tenant participation

There are few references to tenant organisation and activity in the immediate postwar years. However, during the 1950s and 1960s the formation of tenants' associations proceeded gradually although there was a significant increase in the late 1960s and in the early 1970s. A particular impetus at this time was provided by increases in rents and the Conservative government's introduction of 'fair rents' in the public sector, which resulted in a doubling of membership of the Association of London Housing Estates (ALHE). In other places such as Sheffield, the level of tenant organisation and activity grew considerably at this time, and campaigns were mounted including direct action (such as rent strikes) to fight the imposition of higher rents (Lowe, 1986).

In the late 1980s and early 1990s, tenant activity was largely confined to local campaigns and issues. However, tenant opposition to Tenant's Choice and the widely perceived anti-democratic nature of the voting arrangements provided a focus for campaigning on a national issue. Likewise, tenant criticism of the arrangements for HATs resulted in significant changes to the voting and constitutional provisions by central government These successes indicate the government's concern at least to be seen to be acting on behalf of council tenants.

Of course, great tenant activity does not necessarily mean that there is a growth in arrangements for tenant participation. The first known arguments for tenant participation occurred in the immediate postwar years. For example, a Political Economic Planning tract of 1948 argued for tenant participation. The 1959 Labour Party manifesto proposed a tenants' advisory committee for each local housing authority and this idea was put forward in a series of unsuccessful private members' bills between 1968 and 1971. The first systematic survey of local authority arrangements for tenant participation was published in 1977 by the Department of the Environment (Richardson, 1977). The research found that only 12 per cent of English authorities had 'formal schemes' of tenant participation (including tenant involvement in housing committees or sub-committees, advisory committees and regular discussion meetings between councillors or officers and tenant representatives). If 'irregular discussion meetings' were included the proportion rose to 44 per cent. Over half the formal schemes were to be found in London (Richardson, 1977).

In 1980 local authorities in England and Wales (but not Scotland) were given the statutory obligation to consult tenants on changes in housing management which affect tenants substantially. Our survey of local authorities in Britain showed that, partly as a result of this, the proportion of local authorities with 'formal schemes' had risen from 12 per cent in 1975 to 44 per cent in 1986–7. When informal methods (such as irregular discussion meetings) are included the proportion had risen from 44 per cent in 1975 to 80 per cent in 1986–7. There had clearly been substantial growth in the number of local authorities adopting arrangements for tenant participation. Although most early development was in the major metropolitan authorities, mainly in London, by 1986–7 arrangements had been instituted in many other councils, with the major growth occurring in the non-metropolitan boroughs (Cairncross, Clapham and Goodlad, 1990).

Since 1986–7 there has been further growth. A survey of English local authorities showed that the proportion of councils which held irregular discussion meetings with tenants had increased from 69 per cent in 1986–7 to 82 per cent in 1991. Similarly the proportion of authorities in England with regular discussion meetings had increased from 35 per cent in 1986–7 to 47 per cent, and the proportion with joint advisory committees had increased from 23 to 32 per cent (Bines *et al.*, 1993, p.194).

The growth of arrangements for participation was also reflected in the proportion of local authorities which provided support to tenants' associations. For example, in 1986–7 15 per cent of authorities provided free or subsidised office premises (Cairncross, Clapham and Goodlad, 1990, p.37) and this had increased to 50 per cent in 1991 (Bines *et al.*, 1993, p.193). The same growth has also occurred in Scotland. In 1986–7 35 per cent of authorities held regular discussion meetings with tenants compared to 59 per cent in 1993 (Clapham *et al.*, 1995).

The growth has been stimulated in part by government policy. The Tenant's Choice provisions of the Housing Act 1988 led many local authorities to perceive a threat to their position from tenants supporting other landlords taking over their stock. Therefore, attempts were made to open up communication with tenants in order to improve the service and to increase tenant satisfaction, thus reducing or eliminating any incentive to change. Other programmes have been designed to encourage local authorities to decentralise housing management and to involve tenants in the regeneration of problematic housing estates. The DoE's Estate Action programme, for example, held back some borrowing consent for the renewal of housing areas and made the consent conditional on the DoE being convinced by the local authority that decentralisation and tenant involvement were key elements of the renewal plan. The incentive of the borrowing consent has encouraged many reluctant local authorities to embrace the two elements. Finally, following the government's co-operatives review *Tenants in the Lead* (Department of the Environment, 1990) funds were provided by the DoE (Section 16 funding) to enable tenants' groups to receive support and training in forming a tenant management co-operative or an estate management board. However, some of these policy initiatives from central government designed to increase tenant participation have not been present in Scotland and yet participation has still spread. This could be due to a reaction to the general environment facing council housing, but it also reflects the growing belief among housing managers and councillors that tenant participation is an important element of good housing management.

In summary, tenant participation has been a relatively long-standing phenomenon in British council housing although growth occurred in the 1970s and accelerated during the 1980s and into the 1990s. Tenant participation is now widespread even though only a minority of local authorities have adopted each of the formal mechanisms available.

A variety of factors may account for the growth of tenant participation. Interest in user participation has developed across a range of services provided by the central and local state, from education and health care to social services and psychiatric care. This has coincided with the growth of consumerism in the private sector. Richardson (1983, p.114) attributed the growth of participation to the change in the character of consumers. They had become 'more demanding, more able, more numerous and, not least, more organised'. Participation has become fashionable and is perceived as professional good practice. In the context of party politics tenant participation has achieved all party consensus as a good thing (although definitions of what it means in practice may vary widely). In addition, as public services came under attack in the 1980s, participation was seen as a means of retaining public support.

From the housing managers' point of view, it is perceived as a means of overcoming difficult areas of housing management and providing contact with tenants who have become increasingly remote through local government reorganisations of the 1960s and 1970s and the decline of the rent collector, who used to serve as the landlords' eyes and ears on the estates. Political, professional and social circumstances may all have contributed to the growth of tenant participation in council housing.

CONCLUSION

This chapter has charted the major trends which have occurred in local government in the 1980s and early 1990s. The major trends were continued central government attempts to restrain capital and revenue expenditure of local authorities, although they achieved more success in restraining the former than the latter; the reorganisation of local government and its fragmentation into a large number of different agencies; the privatisation of some local authority assets and services; growing commercialism in the management of local government services; increasing central government involvement in areas which have traditionally been the preserve of local government; and

21

finally the increasing involvement of users in the provision of services. All of these trends were present in the management of public rented housing and have led to changes in the role of council housing and the way it is run. Of particular importance to our focus in this book is the growth in the arrangements for tenant participation.

Our review shows that the restructuring of public housing cannot be characterised simply as a set of central government policies which have been imposed on protesting local authorities. Some of the changes, most notably the drives towards tenant participation and the decentralisation of housing management, serve to reinforce existing moves in these directions by some local authorities of varying political persuasions, although predominantly under Labour control. Community Ownership in Scotland is one example of this where central government adopted a programme first developed by Labour-controlled Glasgow District Council.

The 1980s and early 1990s have been a period of considerable change for public rented housing. Central government has been successful in its attempts to reduce capital expenditure on council housing and to reduce central and local revenue subsidies. There has been a growing fragmentation of the provision of public rented housing with the role of councils as providers in decline and housing associations, tenant management co-operatives and other forms of rented housing on the increase. Privatisation has taken place, most notably through the sale of council housing, but also through measures such as the use of loans from private financial institutions in housing association developments and the introduction of CCT into housing management. Increasingly, the market and its values have penetrated the provision of rented housing, shown most clearly by reforms in the housing association sector, but also by CCT and other changes in council housing management such as the increasing use of performance measurement. Increased central government involvement in the management of public rented housing took place through the growth of centrally regulated housing associations and through the increasing use of earmarked funds such as the Estate Action programme to encourage local authorities to adopt central government priorities. There has also been increasing involvement of tenants as users of the service in the management of public rented housing through an increase in co-operatives and in other tenant participation initiatives.

The impact of these changes has varied considerably in different parts of the country. In some relatively rural areas mostly in the south of England, it has meant the complete removal of council housing through the combined impact of the Right to Buy and voluntary transfer to a housing association. In other English rural areas the scale of council housing has reduced considerably through the Right to Buy. This has also had some impact in urban areas, but there are still large stocks of council housing in the major cities such as London, Sheffield, Glasgow and Birmingham. The effect of these measures has been to change the context within which the management of council housing takes place. As we shall show in this book, the result, in some local authorities at least, has been to change the way that council housing is managed, by altering the relationship between housing managers, councillors and tenants.

These trends cannot be understood as merely the results of central government action. Every action of central government provokes a reaction from local authorities and, as we have shown, many of the changes, such as decentralisation or user involvement, were pioneered by local authorities before central government adopted them and sought to promote them through various means. Some of the actions of central government, such as the restrictions on capital expenditure, have been bitterly opposed by some professionals and councillors in local government. Others, such as the involvement of users, have been widely supported. In other cases, such as the Right to Buy, there has been initial hostility which has turned to grudging acceptance once the popularity of the measure has been demonstrated. However, even where there is agreement on particular programmes between local and central government, the parties may have different objectives.

Conflicts such as this draw attention to the battle of ideas which is taking place around local government. There are clearly different views of the role of local government and the way that services are provided which underlie many of the changes and conflicts outlined here. It is to these underlying ideas that we turn in the next chapter.

ABOUT THIS BOOK

The next chapter completes the context-setting by concentrating on the battle of ideas concerning the management of local government, particularly focusing on the debate over ideas of consumerism and citizenship. Our argument is that tenant participation in its different guises reflects the application to the management of council housing of ideas which have their roots in different political philosophies.

We identify three ideal types of local authority which differ in their relationships between housing managers, councillors and tenants. The first type is the traditionalist authority which has only introduced limited tenant involvement, preserving as far as possible traditional patterns of representative democracy and professionalism. The second type is the consumerist authority which has adopted many private sector business practices. Tenants are viewed as individual consumers and the emphasis is on a two way flow of information between landlords and 'consumers' and the exercise of consumer choice. The third type is the citizenship authority which lays stress on the role of local people as being more than just consumers of services, placing emphasis on the collective involvement of tenants and on their rights and obligations as citizens.

Part II of the book is in four chapters which concentrate on the actors in the participation game in the different types of authority. Chapter 3 looks at housing managers, examining in particular their attitudes towards professionalism and the impact these have on the perceptions of the role of the other parties. Chapter 4 focuses on the views and attitudes of councillors towards their role and that of the other parties. It compares these views with traditional typologies of local councillors and suggests that these need some amendment in the light of evidence about councillors' attitudes towards the idea of participatory democracy.

The role of tenants is dealt with in two chapters. Chapter 5 deals with tenants as individuals concentrating on their attitudes towards their landlord and the housing management services provided. Also, it looks at tenants' views of the tenant participation policies and practices of their landlord. The chapter shows that there is substantial dissatisfaction with the services provided and the landlords' treatment of tenants. In all types of authority there are complaints of paternalism from housing managers and councillors and a demand for tenants to have more say.

Chapter 6 focuses on the collective representation of tenants through tenants' associations. The structure and activities of associations are outlined showing that many of them are small and live a fragile, hand-to-mouth existence. The motives of tenants in joining and becoming active in a tenants' association are analysed and the views of tenant activists towards councillors and housing managers are outlined. It is argued that the close links many tenant activists have with their landlord leads them to have a more favourable attitude towards them than tenants as a whole. However, this can put some activists in a difficult position between ordinary tenants and the landlord.

Chapters 7 and 8 constitute the final part of the book. Chapter 7 provides a framework for analysing the relationships between councillors, housing managers and tenants. Using a theoretical conception of power derived from Clegg (1989) the main factors which influence the relationships are illustrated. Clegg views power as being inherent in contingent and ever-changing relationships rather than as a possession which individuals or groups can hold. Therefore, there is no clear answer to the main question posed by the chapter, which is whether the power of tenants differs between the three types of authority. Nevertheless, some insight is gained into the factors which help or hinder tenants.

Finally, Chapter 8 reviews the evidence on the relationships involved in the traditional, consumer and citizenship authorities. In particular, it examines the position of tenants in each of these situations and argues that none of them in practice has overcome tenant dissatisfaction or the traditional relationship between landlord and tenant. The future of the three types of authority is assessed and it is argued that citizenship authorities may have great difficulty in continuing as at present, whereas the consumerist approach could be applied by any public or quasi-public landlord. Future changes could be quite radical as the models change under the influence of outside pressures and changing relationships between councillors, housing managers and tenants.

The reforms described here have been the focus of considerable political conflict between political parties and between central and local government. At the heart of these conflicts have been different ideological views of the scope and nature of public services. Therefore this chapter is closely related to the discussion in Chapter 2 of the ideologies underpinning the changes.

2

CONSUMERS OR
CITIZENS?

In the last twenty years there have been sharply contrasting views
of the role of local authorities, and their relationship with the
electorate, as witnessed by the changes which have taken place and
the conflicts they have engendered. The aim of this chapter is to
outline the major ideas involved, before concentrating on their
impact on the relationship between tenants, councillors and housing
managers. Among the two major ideas which have dominated
thinking about local government are consumerism and citizenship.
This chapter develops three models of local authorities based on
these concepts and looks at what they mean for the relationship
between landlord and tenant in general and tenant participation in
particular. This discussion is illustrated with details of our six case
study local authorities.

THE BATTLE OF IDEAS

Many people in local government still hold what we call 'traditional'
ideas about the roles of councillors, officers and the electorate. Such
people operate within a set of assumptions about the professional
role of the housing manager and the authoritative position of local
authority councillors in representing the views and wishes of the
local electorate, including council tenants. This view has been challenged by the two concepts of consumerism and citizenship, both
of which involve changes in the roles of the actors involved in local
authority housing management. The policy changes of the 1980s
and 1990s discussed in Chapter 1 partly reflect the development of
these ideas at national and local level. On the basis of the ideas it
is possible to identify three models or ideal types of local authority:
traditionalism, consumerism and citizenship.

Table 2.1 Models of local authority roles

	Traditionalism	Consumerism	Citizenship
Primary focus	Focus on producers, i.e. housing managers and councillors	Focus on role of consumer	Focus on consumer and citizen
View of tenants	Focus on needs of tenants as a whole	Focus on individual tenant	Focus on tenants as individuals and as a collective
	Paternalism and authoritarianism	Emphasis on tenant choice	Tenants' rights and obligations
Information flows	Reliance on political and professional judgements	Market research	Dialogue
	Information transmitted through professional and formal political channels, i.e. ballot box	Advertising	Two-way information flow through many channels
Issue focus	Focus on general issues relating to tenants as a whole	Focus on issues directly relevant to individual tenant	Focus on individual and collective issues

The three approaches are contrasted with each other in Table 2.1 and are then considered in turn. Of the three models, consumerism is the one which is closest to the government's vision during the 1980s and early 1990s and underlies many of the changes in local government and housing outlined in the last chapter.

THE TRADITIONAL MODEL

The traditional model is built on the ideas of representative democracy and professionalism. Traditional professions such as medicine have a body of knowledge to which others are denied access and which is held to be universally appropriate. Doctors have traditionally viewed patients as their clients, although they are increasingly being held to account to politicians through the National Health Service. Within local government a similar approach is taken by the newer

27

profession of social work which views users as 'clients' but is directly accountable to local politicians who, in theory at least, determine priorities and allocate resources. The housing profession has not developed the focus on individual clients as in social work or medicine, and neither has it in the past developed forms of accountability to tenants collectively. Therefore, local councillors have largely been seen as representing the clients.

Traditionalism is also built on conceptions of representative democracy with elected politicians acting as policymakers and as the key communication channel between the council and the tenants. Tenants' groups may seek to influence the councillor by acting as a pressure group, but the primary channel of influence is through elections and through individual contact between members and constituents, for example, in surgeries. The relationship between the housing authority and tenants is paternalistic and possibly authoritarian. Councillors and professionals act with their interpretation of tenants' interests at heart, but believe that they are best qualified to decide and act on behalf of tenants.

THE CONSUMERIST MODEL

The idea of consumerism stretches back to the 1960s when a consumer movement associated with activists such as Ralph Nader campaigned for more consumer rights. However, the concept of consumerism which has assumed great importance in local government in the last few years, like most influential management ideas, has its roots in the private sector and seems to stem from the study of successful American corporations carried out by Peters and Waterman (1982). They argue that although many companies claimed to be in touch with their market, in many cases the customer was either ignored or regarded as a nuisance. Excellent companies, in contrast, were obsessed with quality, with customer service and with listening to users. The idea of 'sticking close to the customer' caught on in private industry in the USA and Britain and spread to local government in those countries (Hambleton, 1988). In Britain the National Consumer Council (1986) produced a guide to local authorities on how to carry out and use consumer assessments of services. The Local Government Training Board (1987) produced *Getting Close to the Public* which was intended as a handbook for councillors and officers, outlining consumerist ideas and offering guidance as to how they could be implemented in local government.

Consumerism in the public services struck a chord with politicians influenced by 'New Right' thinking. This drew on the work of public choice theorists such as Buchanan and Tullock (1962) and Niskanen (1971; 1973). They argued that public organisations had an inbuilt tendency to grow in size and to overproduce. This was held to be partly because of the perceived motivation of senior bureaucrats to gain prestige, higher salaries and security by expanding the size of their empires. It was also said to be because of the system of representative democracy which, they argued, led to politicians responding to pressure group demands for more spending. Political parties were said to vie with each other to spend more money and offer more services in order to woo the electorate while hiding the financial consequences of their commitments. In any case existing democratic arrangements were held to be poor mechanisms for meeting the preferences of citizens, due to inadequate voting mechanisms offering voters only one chance every few years to choose between a few packages of policies covering a whole range of issues.

The inadequacies of the political system were contrasted by public choice theorists with the perceived success of market mechanisms in providing choice and responding to consumer preferences. Therefore, the public choice approach provided support for the attempts by central government in Britain in the 1980s and early 1990s to restrain local government expenditure, to privatise where possible and to fragment and commercialise what was left. Consumerism was seen as an appropriate ethos for a more market oriented local government. In the 'consumerism model', emphasis is placed on the individual household either to begin or continue to consume its product in preference to a competitor's. The emphasis is on ensuring that the household can gain easy access to the product or service; providing information in a way which encourages take-up; promoting a positive image (advertising); and providing the consumer with a choice of product or service and giving them a means of redress if they have complaints. Also, the producer needs to keep in touch with the changing needs, demands and aspirations of consumers through market research. The influence of these ideas is reflected in the finding that in 1990–1 over half (56 per cent) of England's housing authorities completed a survey of tenant satisfaction (Bines *et al.*, 1993, p.73).

THE CITIZENSHIP MODEL

There was some resistance to the introduction of consumerist ideas by defenders of local government. Despite a strong belief in local government providing collective services, it was argued that local authorities needed to reform themselves and adopt a new approach. For example, Stewart and Clarke (1987) argued that consumerism by itself was not sufficient as a guide to action in local government. They sought to combine consumerism with the concept of citizenship to form a 'public service orientation' which they argued could provide an organisational culture which was appropriate for the pressures and obligations facing British local authorities.

The attempt to combine consumerism and citizenship stems from arguments that the consumerist model does not entirely fit the normal mode of public service provision. For example, Hambleton (1988) has argued that consumers of public services are not in the same position as consumers of private services because they do not always have the choice of shopping elsewhere. He argues that individual consumers of private services have more choice than consumers of traditional local government services even when they are privatised. 'Individuals cannot switch their refuse collector, their fire service, their water company, their electricity board, etc.' (Hambleton, 1988, p.129).

Stewart and Clarke (1987) also emphasise the distinguishing features of the public sector: the customer does not always buy the service; sometimes the customer may be compelled to receive the service; allocation criteria based on need rather than demand are sometimes used; and limits on resources may well determine the amount of the service available. In addition, the relationship between consumer and provider may itself be a key part of the service, as in counselling. Therefore, the relationship between public producers and their customers may not be as straightforward as that between private producers and their customers.

As many public services are provided and consumed collectively, it is arguable that the individualism of 'consumerism' is not an appropriate way of expanding choice and ensuring consumer power. Hambleton argues that collective services can only be influenced by consumer power through the exercise of political control. Thus, 'we need improved mechanisms for enabling different groups of consumers and citizens to influence political decision-making about the *collective* provision of services' (Hambleton, 1988, pp.129–30). However, services such as housing management have both individual

30

and collective elements. It is true that there are some aspects of management which provide a collective benefit – such as estate management – and it is true that the option for individual tenants to change their housing manager is limited. Nevertheless, some aspects of housing management can be carried out on an individual basis as the example of owner-occupation shows. However, this tenure also demonstrates the nature of collective decision-making, as in the national framework of tax advantages, and at the local level owner-occupiers in flats need to band together in order to deal with the communal aspects of their housing. This implies the desirability of decentralisation of political power to consumers acting collectively, strengthening the citizenship approach.

Despite reservations about the concept of consumerism, Hambleton concludes that local authorities have much to gain from importing consumerist ideas and adapting them to public service needs, but he argues they should address the question of empowering users which will not happen through consumerism alone. Winkler (1988) takes up this point in her criticism of consumerism in the health service. She argues that consumerism was about customer relations not patients' rights, 'this model is a "harmless version" of consumerism – it requires little serious change, but much public visibility. It is about the appearance, not substance of change' (Winkler, 1988, p.1).

Stewart and Clarke's (1987) attempt to adapt consumerism by combining it with citizenship raises a number of dilemmas. In particular, they accept that local authorities have to consider not only the users of the service, but also the public at large to whom they are democratically accountable. The emphasis is both on the *customer* for whom the service is provided and on the *citizen* to whom the local authority is accountable (Stewart and Clarke, 1987, p.169). There is a clear potential for tension in this approach, but Stewart and Clarke do not discuss how tensions between customers and citizens should be resolved.

Rhodes (1987) is quite clear that local authorities are primarily political organisations. He suggests that the 'public service orientation' should be defined as '"the 3cs": consumerism, caring and citizenship and . . . the greatest of these is citizenship' (Rhodes, 1987, p.64). The implication of this type of argument is that consumerism should be allied to an enlargement of the political accountability of local government. Therefore, as well as having the rights of individual consumers, tenants collectively should be able to participate in the provision of the service. As citizens, tenants have obligations as well

31

as rights, most notably the obligation to consider the interests of prospective tenants, other tenants and non-tenants.

In consumerism, the relationship between the producer and consumer of housing management services is viewed in the same way as in any other market relation. The citizenship approach has five major differences. It recognises first that the views of citizens in general as well as consumers should be considered. The second emphasis is on collective, as well as individual involvement. Third, there is held to be an obligation on public organisations to treat users as both consumers and citizens. Fourth, the focus is on consumer and citizen rights and obligations as well as on consumer choice. A fifth key element of the citizenship approach is the existence of a dialogue between the producer, consumer and citizen which is more than the two way flow of information involved in consumerism, but involves negotiation and trade offs.

TENANT PARTICIPATION

The three models or ideal types of the traditional, consumerist and citizenship local authorities can be applied to the relationship between councils and their tenants as mediated through the mechanisms of tenant participation. As a starting point it is necessary to make a definition of tenant participation.

A definition of tenant participation

Most attempts to define tenant participation have involved the construction of one all-inclusive definition. For example, the Institute of Housing and the Tenant Participation Advisory Service have defined tenant participation as:

'A two-way process involving sharing of information and ideas, where tenants are able to influence decisions and take part in what is happening.'
(Cairncross, Clapham and Goodlad, 1989, p.19)

The problem with this approach is that simple, all-inclusive definitions hide important differences between various forms of tenant participation. Therefore, we have disaggregated tenant participation into three elements: structures, processes and objectives (Cairncross, Clapham and Goodlad, 1989). Using these elements it is possible to construct definitions of different kinds of participation.

32

Each of the three elements is divided into a number of categories. For example, Table 2.2 shows different structures or methods of communication between tenants and their landlord. To this can be added the questions or issues which are subject to participation, such as modernisation of the housing stock or the extent of participation arrangements, such as whether they involve tenants living in all areas, or are restricted to particular estates.

The second element is the process which occurs through the participation arrangements. A number of processes are listed in Table 2.3. These are in no particular order and do not constitute any form of 'ladder' of participation. We hold that it is impossible to construct such a ladder as the processes vary on more than one dimension, as will become apparent later.

The third and final element is the objectives of participation. Five aims, identified in Figure 2.1, are held by participants such as councillors, housing managers and tenants. In almost all cases tenants have only one reason for getting involved in tenant participation, which is to improve their living conditions. In other words, they adopt an instrumental view. Although tenants' representatives may gain in terms of personal skills from the experience, this is not why they become involved.

Disaggregation of tenant participation in this way allows more insight into the battle of ideas involved in participation. Using clusters of different structures, processes and objectives, three ideal types can be identified which align with the general models of traditionalism, consumerism and citizenship outlined earlier. It must be emphasised that these are ideal types and few local authorities conform exactly to them. However, as we shall show, the models illustrate what is happening in council housing. The ideal types have been partly derived from the abstract application of the general ideas outlined earlier and partly from observing tenant participation in practice in six authorities, and from analysis of surveys of tenant participation structures.

The six case study local authorities were selected to represent different experiences of tenant participation, and to reflect differences in size, location and political make-up. Two had no formal arrangements for participation and no known tenants' association (Talybont and Braehead); two had no formal arrangements for tenant participation and yet had a significant number of tenants' associations (Dockport and Sainsborough); and a third pair had well-developed, formal arrangements for tenant participation and a significant

Table 2.2 Methods and structures for tenant participation

Written communication	*Face-to-face contact*
Can take various forms of which the most common appear to be:	Can involve contact with one tenant, but usually involves greater numbers as at a public meeting, or tenants' representatives, either from a tenants' group or specially selected/elected. Methods include:

• letters
• leaflets
• handbooks/information packs
• advertisements
• posters/notices
• newsletters
• questionnaire surveys
• exhibitions
	• tenant representation on housing/management committee
	• tenant representation on sub-committees
	• tenant representation on working parties or advisory committees (not part of the decision-making apparatus)
	• meetings between housing authority/association representatives and tenants
	  • at regular intervals
	  • at irregular intervals

Community development
Tenant participation plays a part in improving individual tenants'
skills and capacities which in turn leads to a stronger sense of community and to a reduction in social problems. Tenants are enabled to
play a greater part in determining the future of their area and to be
more effective participants. Tenant participation is thus seen as a
valuable activity in its own right.

Better housing and housing management
Tenant participation assists better decision-making by providing
information which can be taken into account in service delivery and
planning. This can lead to better decisions and more satisfied tenants.
Participation is, therefore, a means to an end.

More choice or power to tenants
Participation can be seen as essentially bound up with the moral
right of tenants to influence their own living conditions, either on
the basis that the tenant pays or because of the implicit view about
the nature of a democratic society.

Tenant satisfaction
Participation plays an important part in effective housing management, in which effectiveness is strongly associated with tenant satisfaction. Participation is, therefore, a means to the end of achieving
more satisfied tenants.

Helping councillors or committee members
Tenant participation can be seen as directly in conflict with the roles
of councillors and committee members, but an increasingly common
view is that participation of tenants can complement and supple-
ment the role of policy makers. Tenants are seen as providing useful
information to committee members who can assist the tenants to
present their point of view, especially when something appears to
have gone wrong with the system for handling complaints. Tenant
participation can, therefore, assist policy makers to be more effective.

Figure 2.1 Objectives of actors in tenant participation

number of tenants' associations (Strathmuir and Compton). (The
authorities have been given fictional names to preserve confiden-
tiality.) Following the fieldwork, the authorities were reclassified using
the typology discussed in this chapter. Talybont, Braehead and
Dockport are defined as broadly matching the traditional ideal
type; Sainsborough is a consumerist authority; and Strathmuir and
Compton conform reasonably closely to the citizenship ideal type.

Table 2.3 Processes of tenant participation

Providing information	not a form of participation in itself but an essential prerequisite. Can be proactive or reactive. The written and spoken word are the major media for information provision.
Seeking information	arguably also a prerequisite rather than participation itself. Here the landlord seeks information about facts and opinions, using methods such as surveys, meetings and letters.
Listening	takes place if landlords listen to the views of tenants expressed verbally or in writing, and can take place at a distance and without the tenants knowing their views are being listened to.
Consultation	overlaps with seeking information and listening. The emphasis is on asking for views in order to consider them before reaching decisions, with sufficient time for tenants to formulate their views and landlords to consider them. Face-to-face contact is not necessary though may be desirable.
Dialogue	involves all the previous forms or processes of participation, but unlike them is likely to involve only tenants' representatives and groups, rather than tenants individually. Most of the common structures intended to achieve dialogue are established for a period of months or years. Dialogue allows for negotiation to take place in relation to specific issues, and implies that both parties have an interest in reaching a mutually satisfactory conclusion. In other words the landlord must want the tenant(s) to accept the outcome and vice versa. This is more likely to be the case where the tenant (or landlord) has some sanction to use against the other in the event of an unsatisfactory conclusion.
Joint management	provides tenants with a decision-taking role along with housing authority/association representatives.
Choice	allows tenants to choose from alternatives agreed by the landlord and can be in an individual or collective capacity.
Control	not the only 'real' form of participation, and arguably not a form of 'taking part' at all in that tenants are enabled to take over and directly run certain aspects of their housing services (within the law).

TRADITIONAL TENANT PARTICIPATION

The first ideal type is that of traditionalism. Local authorities in this category have changed marginally their methods of operation by embracing only limited aspects of tenant participation as shown in Table 2.4. In general such local authorities do not adopt face-to-face structures of participation but rely on providing information to tenants through written material. In England and Wales the statutory requirement to consult tenants is met through sending letters to individual tenants informing them of changes in housing management and asking for a written, individual response. This was the situation in Talybont, a small Welsh authority, and in Dockport, a metropolitan borough in the north of England. In Braehead, a medium-sized authority in central Scotland, the provision of written information for tenants had increased in the 1980s, and remained the main example of 'participation'.

Local authorities in this category do not seek information from tenants in general, although some may commission a household survey of tenants as a means of gauging satisfaction with the service provided. None of the three traditional authorities – Talybont, Dockport and Braehead – had formal policies on tenant participation, and only one, Talybont, had carried out a satisfaction survey. Much reliance appeared to be placed on traditional notions of representative democracy and professional expertise, but some attitudes appeared to be changing, partly as a result of Tenant's Choice.

Most participation in traditional local authorities will revolve around modernisation and improvement programmes and difficult-

Table 2.4 Traditionalist tenant participation

Structures	Processes	Objectives (councillors, officers)
Letters	Information provision	Better housing management
Leaflets	Listening	Tenant satisfaction
Handbooks	Seeking information	Community development
Survey	Choice	(tied up to previous two)
Most participation around:		
Modernisation		(Tenants)
DTL estates		Better housing

to-let estates. For example, in Braehead's modernisation and improvement programmes there were usually public meetings to explain the programme, and individual visits to tenants, as well as written methods of communication. The council listened to the views of tenants, although it was too late to change the overall modernisation package. Individual tenants were, however, given some choice over details of the work carried out. For example, they may be able to choose the colour of their bathroom suite or the type of fuel for the central heating system.

In difficult-to-let estates, structures of participation may be more extensive. Where no tenants' association exists there may be attempts to set one up, as in Dockport. Regular meetings may be held with groups of tenants in which the local authority may seek out the views of tenants as to how to solve problems in the area. In Braehead's most problematic estate relationships between the tenants' association and the housing department improved markedly after a housing office opened in the estate, headed by an officer recruited from a larger authority with greater experience of tenant participation.

In traditional authorities there will be few if any tenants' associations and little tradition or experience of them. In Talybont, Dockport and Braehead the local authorities were not a very reliable source of information about the small number of tenants' associations which were found to exist. Dockport had ten associations, and Talybont and Braehead both had three or four active during our fieldwork. Outside of some difficult-to-let areas there was little or no help for tenants' associations through, for example, the provision of grants or help in kind such as free access to office facilities. Among tenants there was a low awareness of tenant participation options.

The ultimate aim of traditional local authorities is to create high tenant satisfaction primarily as a means of ensuring, after the 1988 Housing Act and other changes in council housing, that tenants do not transfer to another landlord:

'The ultimate objective is to make sure they stay with us.'
(Housing Manager)

The emphasis on the provision of information is designed to help explain to tenants why certain decisions are taken and to improve the image of the local authority.

A subsidiary aim, closely connected to the desire to increase tenant satisfaction, is to improve housing management. This objective is also seen as a way of easing the problems that housing managers and housing authorities face, particularly in difficult-to-let estates.

'You cannot manage property unless the people whose homes we are managing are with us. They are their homes and they know a great deal more than we do. Even as officers we may think an issue is crystal clear and obvious, but it may be wrong because we haven't thought about all the angles. They [Tenants' Associations] can bring things to our attention which we may not be aware of. They can help us form our policies in such a way that they're going to be successful.'

(District Housing Manager, Dockport)

Traditional local authorities have adopted an essentially defensive and reactive stance to the changes in policy towards council housing. Tenant participation is seen as a mechanism to prevent the loss of stock to other landlords and as a way of easing some of the burdens on housing managers and councillors caused by the overall physical and social deterioration of council housing. One councillor in Braehead, for example, felt that tenant participation would not be 'required' if the government would allow the council to spend more to meet housing needs.

This form of tenant participation reproduces existing power relations because little influence is achieved by tenants. Communication is primarily from the housing authority to tenants. Where communication is in the reverse direction, it largely follows an agenda established by the local authority which usually reserves the right to make all decisions except over the most detailed and practical of issues (such as colour of bathroom suites). This does not mean that tenants cannot gain some influence in 'traditional' local authorities. However, any influence is likely to be limited in extent and duration.

There is little ideological basis to the attitudes adopted in such authorities apart from a belief in traditional structures of representative democracy and of professional competence. As a consequence traditional local authorities can be found which are controlled by either of the major political parties or by any combination of parties in a 'hung' council. Braehead was solidly Labour; in Dockport the Conservatives had recently lost overall control but remained the largest party; and in Talybont there was a Labour administration based on the urban areas, with strong Conservative and independent representation in the rural areas. In all three authorities a small number of New Left Labour and minority party councillors

Table 2.5 Consumerist tenant participation

Structures	Processes	Objectives (councillors, officers)
Letters	Providing information	Better housing
Leaflets	Seeking information	management
Handbooks/	Listening (to individual)	Expand choice
information packs	Choice	Tenant satisfaction
Advertisements/		
newsletters		(Tenants)
Surveys		Better housing
Meetings with		
individual tenants		
Issues which impact		
on individual tenants		
rather than general		
policy		

demonstrated attitudes more characteristic of consumerist and citizenship councils, as did a few officers.

CONSUMERIST TENANT PARTICIPATION

The second approach by local authorities we have called consumerism because of the desire to encourage the view of rented housing as a commodity and to institute a more commercial relationship between landlord and tenants. The main features of the consumerist model applied to tenant participation in council housing are given in Table 2.5.

Consumerist local authorities place a lot of emphasis on the presentation of information to tenants in order to improve their image. As well as leaflets and handbooks they may produce newsletters and may advertise their services in the local media. They also give a lot of attention to 'market research' or gauging the views of their customers (tenants) on the service received through surveys or through soliciting responses from users of particular services such as repairs or allocations. Also, a lot of attention is focused on meetings with individual tenants, usually over the counter in a housing office.

Sainsborough, a hung council in the south east of England, used postal and door to door surveys to collect tenants' views, and devoted more resources to information provision than the traditional authorities.

However, a consumerist local authority may not provide any support for tenants' associations and may refuse to meet with tenants' representatives on the grounds that communication should be with individual tenants. The authority may question the legitimacy of tenants' groups and could at best ignore them and at worst actively obstruct them. Therefore, there may be few tenants' associations and those that exist are likely either to be primarily social organisations or oriented towards opposition to the local authority. In Sainsborough, for example, many tenants' associations had existed in the 1970s as the authority, then Labour-controlled, pioneered tenant involvement on district and authority-wide committees. But, after a period of New Right Conservative control which ended just before our field work began, only seven associations remained, and relationships were now characterised by mutual caution. As in the traditional authorities, most encouragement of tenants' associations took place in areas undergoing modernisation.

Consumerist local authorities restrict the issues on which participation takes place to those which impact on tenants individually. Therefore, they may canvass views on a particular estate management issue from those individuals affected by it, or may give individuals choice over when a repair is carried out, or the colour of their decoration. There is little or no participation over issues of general policy such as rent levels, allocation policy and so on. It is felt to be right that these issues are decided by the landlord in the light of information about consumer preferences gained through surveys and from consumer behaviour. In Sainsborough, for example, one district manager criticised a tenants' association for overstepping its remit when it objected to the authority subsidising the General Fund from the Housing Revenue Account.

The emphasis of local authorities following the consumerist model is on providing information (advertising and public relations) and seeking information (market research). They are likely to listen to individual tenants but not to any tenant organisation. Perhaps above all, the individual tenant is seen as participating in the management of housing by exercising choice in the same way as a consumer in a market. Senior officers in Sainsborough tended to identify tenant participation with professional housing management, seeing it as an aid to providing a good quality service with opportunities for customer feedback. Advocates of this approach in local authorities share the objectives of the traditionals of creating tenant satisfaction and improving housing management. However, they tend to see these objectives in market terms and express them in commercial language. The best exposition of this view came from a consumerist housing officer in a traditional authority.

'I am a professional housing person and I have a quite deep-rooted desire to offer a good service, and the only way you can do this is to . . . the only way the customer will think it is a good service is to give them what they want.'

(Housing Manager, Talybont)

Councillors in consumerist local authorities are likely to see the creation of market or quasi-market relations and the expansion of consumer choice as an end in itself which conforms to their ideologically based view of the appropriateness of a market model of housing.

In summary, consumerist local authorities pursue an individualist form of tenant participation as part of the establishment of market relations in council housing. Tenants are seen as customers and empowering tenants is viewed as being achieved through giving them the role of a consumer who is able to exercise market choice. Rented housing is viewed as a commodity like any other, and the council produces a housing service in much the same way as any other service producer such as a supermarket. There is no place in this approach for tenants to act collectively and this form of activity will be resisted or ignored.

Not surprisingly consumerist local authorities tend to be Conservative-controlled. In Sainsborough consumerism was taken further by the initiation of a process – later abandoned – of transferring the stock from the local authority to a housing association. There is a consistent logic to this transfer. If the landlord–tenant relationship is primarily one of service provider and consumer, there is no reason why the service should necessarily be provided by local government. This is an approach which also sits comfortably with the introduction of CCT in housing management.

CITIZENSHIP TENANT PARTICIPATION

The third ideal type can be characterised by the term 'citizenship' as shown in Table 2.6. In this approach local housing authorities use the full range of participation structures ranging from written

Table 2.6 Citizenship tenant participation

Structures	*Processes*	*Objectives* *(councillors, officers)*
Letters	Providing information	Community development
Leaflets	Seeking information	Better housing management
Handbooks	Listening	Empowering tenants
Newsletters	Consultation	Tenant satisfaction
Tenant representation	Dialogue	
Regular meetings	Joint management	(Tenants)
	(Control)	Better housing
Covers whole range of *issues from practical* *to policy*		

materials to tenant representation on decision-making bodies. Support is made available to tenants' associations, which are actively encouraged. There is a system of tenant representation either at the local estate level, where tenants' representatives have a say over the running of their estate, and at the local authority-wide level, or a combination of both. In addition, there are likely to be regular meetings between housing managers and tenants' associations at all levels outside the formal structures. The balance between these three elements (local representation, authority-wide representation and trades union style bargaining through regular meetings) may vary considerably between local authorities. Nevertheless, all three elements are likely to be present to some degree, as they were in Strathmuir, a large Scottish authority, and Compton, an inner city authority in the south east of England.

Tenant participation may cover a wide range of topics from day-to-day issues affecting particular groups of houses, to policy issues which have an impact on all tenants, such as rent levels and structures or the terms of the tenancy agreement. In these local authorities there will probably be a long history of tenant organisation with many tenants' groups organised into federations at the local authority and perhaps at other levels. Support will be provided by the local authority to tenants' organisations usually in the form of financial help, meeting rooms or halls and access to office facilities. The housing department will employ specialist tenant liaison or community development staff. Among tenant activists there will be an expectation of involvement in decisions. Some councillors will have emerged from the tenants' movement. In addition, some tenants' leaders will have had experience in the trades union movement which may also have a long tradition in the area. All of these features were present in Strathmuir and Compton, both strong Labour authorities.

The wide range of structures used is reflected in the variety of processes which take place. As well as the provision and seeking of information, tenants are consulted over policy and practical issues and there may be a dialogue between landlord and tenant with negotiations taking place before a final decision is reached. For example, in Strathmuir and in Compton the tenants' federations had negotiated the terms of new tenancy agreements with the councils. In addition, there may be structures of joint management either at the estate-based level or at the authority-wide level in which responsibility for decision-making is shared. For example, there may be tenant representatives on council committees (although the scope for this has been narrowed by recent legislation). There are also likely to be tenant management co-operatives or other forms of tenant control on estates, as in Compton and Strathmuir. The two authorities differed, however, in the extent to which they relied on a formal structure of area or district committees with tenant representation. In Compton the main mechanism for participation was through tenants' associations which were able to send their representatives to district committees which in turn were able to nominate tenants' representatives to the housing committee. Tenants' associations that did not participate in district committees could be disadvantaged relative to those that did. In Strathmuir a system of area sub-committees existed but each was free to co-opt community representatives as it wished, so some had no tenant representation. The structures for participation in Strathmuir, therefore, varied from one part of the authority to another, and had varied over time to include and then exclude tenants' representatives as voting members of sub-committees. Councillors argued it was more appropriate to involve tenants in other ways such as policy development working parties.

As well as the objectives of tenant satisfaction, community development and better housing management, there is also likely to be a desire by councillors and housing managers to empower tenants in citizenship authorities. For example, one local authority community worker gave the following justification for tenant participation:

'Because individuals collectively have a little bit of power, individually they have none and they have the common problems of living in poorly run, managed, paternalistic housing and to get any concessions at all out of the council, just for basic repairs services or whatever, they need strong effective organisations and that is the purpose of tenants' associations or should be.'

(Community Worker, Compton)

This kind of justification is often given for the support which is provided to tenants' associations. Councillors in citizenship local authorities may have an ideological commitment to sharing power with tenants collectively.

'To my mind it should be a partnership. The tenants outside with the housing department and members trying to work together as a team to provide better housing, better repairs and a better environment to live in.'

(Councillor, Strathmuir)

In summary, in the citizenship model the focus is on collective representation in sharp contrast to the individualist orientation of the 'consumerist' model. This does not mean that citizenship authorities do not pay attention to factors such as individual choice or the provision of information to individual tenants. Rather, there is usually in practice a mix of the two models and many of the features of the individualist approach are seen as complements to, or prerequisites of, a collective approach. Thus, individual tenants need information before they can participate in representative structures of tenants. Individual choice and a user friendly service are not necessarily in contradiction to a citizenship approach. In practice, however, Strathmuir and Compton's tendency to rely on tenants' associations as the main mechanism for dialogue could mean that individual tenants were poorly informed and unaware of the authorities' efforts to encourage tenant participation.

The collective emphasis is held partly because of an ideological commitment by some councillors and officers to collectivism, but also because it is perceived as a more effective method of 'empowering' tenants. The most important process within the citizenship model is that of dialogue and negotiation between tenants and landlords. This involves the landlord being prepared to enter into a dialogue. Two points are important here. First, the willingness of a landlord to listen and respond to tenants can vary considerably between local authorities and even in different circumstances within the same authority. On some occasions a landlord may accede completely to a contradictory tenant view expressed by representatives; on other occasions or in other places the tenant view may be completely rejected. Many other outcomes are also possible. In practice, it is often difficult to predict how much tenants will achieve, as Chapter 8 argues, but there are usually limits to the formal, devolved, power set by the local authority. For example, some issues may be held to be off the agenda for discussion. In other cases dialogue with tenants' groups is adopted as a strategy to discourage tenants from taking up options of control by forming a co-operative or taking advantage of Tenants' Choice.

Second, the willingness to enter into a dialogue and cede to tenant requests can be withdrawn. This had happened in Sainsborough, where an extensive structure of tenant participation was dismantled by the local authority when political control of the council changed from Labour to Conservative in the late 1970s. Despite protests, tenants' organisations were unable to stop their influence being removed. It is predominantly Labour-controlled councils which adopt a citizenship model.

CONCLUSION

Tenant participation reflects different views of how council housing should be run which have their basis in the political philosophies of market liberalism and social democracy. Therefore, tenant participation in its different guises in the 1980s and early 1990s partly reflects the application of different political philosophies to council housing.

The factor which all the approaches to tenant participation have in common is the belief that some form of involvement of tenants in the management of council housing leads to better housing management. What is meant by 'better' is not always clear but there is a consensus between politicians of different political parties and housing managers that tenant participation is a 'good thing' and should be supported. Nevertheless, the consensus is very shallow as there is extensive disagreement about the form of participation which should be pursued.

Of the three forms of tenant participation outlined here, the 'traditional' approach is essentially an attempt to retain existing power relationships as much as possible in a changing situation. Tenants are involved in housing management to a very limited extent and only in areas where landlords need the co-operation and help of tenants. Therefore, participation is largely restricted to modernisation schemes and difficult-to-let estates. Even here the processes involved are usually ones of information exchange and limited individual choice. In general, adoption of the 'traditional' model presents little challenge to professional autonomy or to the processes of representative democracy.

The other two approaches are based on newer ideological positions. In consumerism, local authorities are intent on treating rented housing as a tradeable commodity which is not the subject of political debate or collective action. Their intention is to introduce the values and methods of the market into council housing by putting landlords and tenants in a producer–consumer relationship. As argued earlier, the logic of this position is either to see council housing as just another option along with housing associations and private landlords, which are more distant from the political system, or to see it removed entirely from the public sector.

The introduction of market relations to council housing in this way has to be seen in relation to concepts of consumerism which have been influential in a wide range of publicly provided services. It must also be seen in the context of market-liberal philosophies which emphasise the importance of market relations. The marketisation of council housing has a symbolic importance because it serves to legitimise these wider philosophies.

A key question is whether tenant participation, as implemented by consumerists, empowers tenants in relation to their landlords. Advocates of the approach would stress the consumer sovereignty which they claim exists in free markets. However, even if this claim is accepted it is debatable whether a free market exists for the council tenant. This issue will be considered in more depth in Chapter 8 after the details of the relationships between tenants and landlords have been considered in the following four chapters.

The third approach of citizenship involves the engagement of landlords in a dialogue with tenants' representatives. The emphasis here is on tenants acting collectively as well as individually. Tenant participation, as implemented in the 'citizenship' approach, is partly a reaction to the position of council housing, but it is also the product of social democratic ideology. Tenant participation serves to illustrate and reinforce concepts of citizenship and to express the importance and legitimacy of the relationship among citizens and between citizens and the state. However, the citizenship model reflects the tension in current social democratic thinking between the tradition of the strong socialist state, and empowerment through collective 'community' control. The 'citizenship' approach is a compromise between these positions.

By disaggregating the elements of tenant participation we have attempted to show that there are different forms which are related to wider political philosophies and reflect the battle of ideas which is currently being fought in council housing. The illustrations from our case studies help to flesh out the three approaches and show how they have been applied in practice. In the following chapters we look in more depth at the implications of the three ideal types identified here for the relationships between tenants, councillors and housing managers.

Part II
THE ACTORS

3

HOUSING MANAGERS

This chapter discusses the role of housing managers, their relationship with tenants, and their attitudes to tenants and tenants' groups. The chapter starts by exploring the concept of professionalism in housing management, which in its traditional form may be expected to reject tenant participation. It continues with an outline of the development of housing management in local government, reviewing the scope of the housing manager's role, and the training and educational background of housing managers. The chapter continues by exploring the attitudes and views of housing managers to tenants, tenants' groups and tenant participation using the evidence from our case studies. Subsequent chapters will show that tenants and councillors often have very different views of the motivations and attitudes of housing managers. The chapter finally draws conclusions about the role of housing managers, the nature of the relationship with tenants and councillors, and how different views of the role of housing managers accord with our framework of ideal type authorities.

Housing managers are often referred to as professionals, and housing management is seen as one of the local government professions. However, the touchstone of professionalism is often seen as being the degree to which occupational groups control their own work. By this criterion, housing managers are weakly professionalised by virtue of their position in local government where councillors and other professional groups such as architects can exert control over their work. Also, the professional body, the Chartered Institute of Housing, does not control entry to the profession and many housing managers are not formally qualified in housing. Therefore, housing managers are on relatively weak ground in arguing that there is a strong and distinct body of professional knowledge (Cole and Furbey, 1994, pp.131–8).

The traditional model of local government sees housing managers as holding expertise and skills which justify their authority in managing the council housing stock with little or no reference to the tenants. Where tenants are 'involved' in management it is largely by having information provided by the professionals rather than by joining in negotiations, consultation or even by providing information to the professionals. The view, formerly taken for granted, of the role of 'professional', expert, housing officer is challenged in a number of ways by the consumerist and citizenship models. In the consumerist model housing managers seek to take account of the views of individual tenants, thus accepting implicitly that 'professionals' do not know everything. In the citizenship model tenants are encouraged to play an active role in influencing the policy and management process, again implying a degree of humility on the part of housing managers which does not accord well with the traditional view of the objective expert. Thus whatever view of their role housing managers hold, it has implications for their approach to their relationship with tenants and tenants' groups.

PROFESSIONALISM IN PUBLIC SERVICES

Social scientists have sought to understand the nature of the relationship between occupational groups and their clients or consumers. The primary concern is with the degree to which occupational groups control their own work, and the extent to which others exercise control over them. Within this approach the defining features of the classic professions such as medicine and the law are the high degree of self government and the strong control exercised over the client. In council housing, the role of councillors and other 'professionals' such as architects or lawyers needs to be considered as well as the role of tenants. Johnson (1972) identifies three types of control operating in the relationships workers have with fellow workers, clients and others such as employers:

1 *Collegiate control,* in which the practitioner defines need and how it is to be met, for example, the nineteenth-century professionals of medicine and the law, and guild control.
2 *Consumer control,* in which the consumer defines need and how it is to be met, for example, patronage in the arts.
3 *Mediative control,* involving a third party, as in the mediating role of capitalists between producers and consumers, and state

mediation such as the role of local authorities in controlling the work of planning officers, and social workers.

Johnson suggests, *inter alia*, that different forms of control are likely to be associated with differing attitudes, beliefs and ideologies on the part of practitioners, and differing social characteristics of the consumers. His typology is perhaps best seen as providing three ideal types.

The first, collegiate control, may be an aspiration for housing managers, and may influence their conduct, but it has little basis in the formal arrangements that apply to the management of most public services, including council housing. Yet housing managers have a set of attitudes to certain aspects of their work, such as allocating vacant dwellings, which appear to owe a great deal to a 'professional' value – allocation according to need – which has not always accorded with councillors' or tenants' views. Despite the low membership of the Chartered Institute of Housing, relative to the number of people employed in housing, there is arguably a community of housing professionals which extends beyond the professional association to encompass all housing managers who are aware of developments in housing policy, who read the specialist housing press or who attend conferences or meetings at which housing issues are discussed. In addition, housing managers are employed (usually) full time, and have greater opportunity than councillors or tenants to acquire knowledge about new developments in housing management thinking. Also career progression leads to movement from one authority to another, possibly transferring knowledge or experience at the same time, as one officer said of his boss:

'One of the big strengths we have is the Chief Housing Officer ... he's very good really. He doesn't meddle over much. He lets staff get on with it, but he's also very experienced, in fact, in some ways, he's probably too good for (Talybont), he worked for the (major urban authorities) and the likes, and he's seen the larger authorities, and he's been in senior positions there. I think he gets a bit sort of uptight sometimes, that we are a small authority and we ought to be doing more than this really.'
(Assistant Chief Officer, Talybont)

So although most aspects of collegiate control are absent from the work of housing managers, there is scope for them to develop a view of how to do their job which owes more to the influence of other housing managers than to the influence of tenants or even councillors to whom they may be formally responsible.

Johnson's second form of control, consumer control, is relevant to the consumerist and citizenship authorities of our typology. But there are some difficulties with it. Neither a consumerist nor a citizenship authority devolves all responsibility for council housing to the tenants. Even in the case of tenant management co-operatives, which give formal responsibility to tenants for running certain aspects of their housing directly, the council retains some control or influence over management functions. Any staff employed by the co-operative, though, can be seen to be controlled by the consumers. But our concern here is with officers employed by local authorities to manage housing. Johnson's consumer control model, therefore, does not take account of the nature of the tripartite relationships between councillors, tenants and housing managers.

Johnson's third category, mediative control, has some parallels with the traditional model of council housing in which officers work within the policy framework set down by councillors, who are accountable to the electorate at local elections. The traditional model lays stress on the professional expertise of the housing manager. Within our traditional ideal type, housing managers are assumed to be trained and equipped to manage a housing service, under the political direction of councillors. Johnson's model suggests that the political control is in fact a mediating influence on the relationship between housing managers and tenants. Equally the relationship between tenants and councillors may be influenced by housing managers – a complexity which Johnson ignores. There is clearly scope here for tension in the relationships between housing managers, tenants and councillors if the political will of the councillors clashes with the professional expertise of the officers. But, of more importance, the exact nature of the mediation provided by councillors or officers is not self evident. Whether housing managers reflect the values and aspirations of tenants, councillors, or housing managers themselves is an open question which is crucial to understanding the nature of the control operating over housing managers.

This chapter is concerned, then, with how housing managers see themselves. Do they see themselves as properly controlled by their peers through a professional ethic, or as servants of their clients (the tenants), or as servants of the council? How does this influence their approach to their relationship with tenants and tenants' organisations? How does mediation operate, over what issues, and with what results? What forms of mediation and control are perceived by housing managers in relation to councillors and tenants? The attitudes of councillors and tenants to housing managers and their view of the nature of their relationship will be addressed in subsequent chapters.

HOUSING MANAGERS AND PROFESSIONALISM

Housing is generally considered to be one of the most important of local government responsibilities, accounting for 22 per cent of gross expenditure (capital and revenue combined) in 1990–1, second only to education (33 per cent) (Department of the Environment, 1992, p.7). Despite this, housing managers do not enjoy a high status in the hierarchy of local government white collar professionals. For example Stewart argues:

> Few people would claim that the profession of housing management had fully established itself on the model of the professions that have traditionally dominated local government. The profession has not yet gained the status conceded to other professions, nor the dominance in the (housing) department characteristic of other departments of the local authority.
>
> (Stewart, 1988, p.39)

There are arguably a number of interrelated reasons for this. One is the slow establishment of housing as a separate occupational category in the local government bureaucracy. Housing departments staffed with specialist housing managers were only slowly established as the stock of council housing built up after 1919. Even today the housing work of local authorities is concerned with influencing the private provision of housing, as well as the management of council housing so town and country planners, environmental health officers and building control officers, for example, have a role in regulating the development or management of housing in the private sector. Within the council housing sector, other professionals, such as architects, have a more influential role in the design of council housing than housing managers: accountants may be responsible for the collection of rents; and engineers may be responsible for the maintenance of the stock (Power, 1987).

A second reason is the perception that housing management is ill-defined as an occupational category. For example, are housing managers more than property managers? Do they have a welfare role?

Does their work encompass other areas of activity such as property development, planning or research? This ambiguity can be seen, it is sometimes argued, when housing is compared with other local government professions such as town planning, environmental health and social work. A third reason for the low professional esteem of housing managers is the generally low educational background of many housing staff. No professional qualification has the same status in housing work as membership of the Royal Town Planning Institute has in town planning, for example.

Stewart offers another explanation:

> The reason may be simple. Housing management may not constitute a profession in the sense established by the older professions. The very use of the word 'management' to describe the profession suggests not a profession, but a particular, management role, requiring specialist skills or knowledge.
>
> (Stewart, 1988, p.39)

Therefore, the relative lack of emphasis on professional skills and training may be due to a smaller emphasis on a professional knowledge base needed to carry out the job. In other words, housing managers may be seen as managers with some professional knowledge and skills rather than professionals with some management knowledge and skills.

Insight into the weak professional status of housing managers in council housing today can be gained through an historical perspective. The earliest group to claim an occupational status as housing managers were the followers of Octavia Hill (1838–1912). The key to Octavia Hill's 'system' was the personal relationship between landlord and tenant (Clapham, 1992b, p.19). Hill attempted to impose her values about thrift and good housekeeping equally on tenants and landlords for whom she acted as agent. She placed emphasis on decentralised management and on putting responsibility for tenant welfare as well as property maintenance in the hands of the housing manager. Her followers were influential in the management of the early housing associations (Brion and Tinker, 1980), but this approach was not generally adopted by local authorities which saw it as expensive, unnecessary and possibly threatening to established professional groups in local government.

The growth of local authority housing from 1919 did not lead to a new professional group taking charge of housing-related work in local government. The housing was managed by officers from a variety of local government backgrounds or disciplines such as public health, surveying or finance – reflecting the variety of departments holding some responsibility for housing development and administration at that time (Power, 1987, Chapter 2). The followers of Octavia Hill set up the Association of Women House Property Managers in 1916, which had numbers drawn largely from the private and voluntary sectors. For many years the two approaches to housing management were represented by different professional associations. The Institute of Housing Administration was set up in 1931, and the qualification at first for entry to this group was holding office at a senior level in local government – this largely restricted entry to men. The two societies eventually amalgamated, in 1965, leading to the Chartered Institute of Housing of today.

The two approaches to housing management represented by the two professional bodies contained different views of the appropriate relationship with tenants. The Octavia Hill tradition was for a personal relationship between housing managers and individual tenants. However, this relationship was not envisaged as one between equals. Hill used her position as housing manager to perform social work on her tenants. She enforced strict codes of moral conduct and would evict tenants who did not live up to them. Housing services, for example having a repair carried out, were used as rewards for appropriate behaviour. Tenants were assessed for housekeeping and moral standards and the best accommodation was reserved for those considered to meet the required standards. Housing management was used as a mechanism for reforming the behaviour of tenants. Her work was more influential on the early housing associations than on local government.

The alternative local authority approach was based, as Power (1987) argues, largely on the management of property rather than people. It was an impersonal form of management based on a written contract and established procedures. The council tenants of the 1920s were relatively prosperous and may have been seen as not requiring moral reform. However, appropriate standards of behaviour were stipulated where it was considered that the property could be misused or damaged, or other tenants could be adversely affected. Therefore, as with the previous approach, the emphasis was on the control of tenants by housing managers.

The two approaches have coexisted in housing management practice throughout the history of council housing, although as we shall see later the approaches have changed considerably as have ideas about the appropriate role of tenants. Nevertheless, as we shall see in Chapter 5, tenants feel that the paternalism inherent in both approaches still characterises the attitudes of housing managers towards their tenants. The balance between the two approaches has changed at different times depending to a large extent on the role of council housing. When the sector has been perceived as catering largely for the poorer sections of the population as in the 1930s or throughout the 1980s and early 1990s, the balance has swung towards the Octavia Hill-originated 'welfare model' (Clapham, 1987). At other times when the sector has been perceived as catering for a wider cross-section of the population, such as during the 1920s or late 1940s, the administrative or property approach has predominated.

Debate over the relevant approach to housing management has been related to discussion of appropriate administrative structures. The 'welfare' approach has usually been associated with a clear view of the coherence of housing management tasks, and the need for a unified housing management structure, whereas the administrative approach has often meant in practice the splitting of functions between several local authority departments.

Power reports that by 1935 'only 13 per cent of the 450 local authorities in Britain had appointed a housing manager' (1987, p.32). In the absence of specialist housing managers the tasks of management tended to be distributed between 'town clerks, treasurers, medical officers of health, engineers and surveyors' (Central Housing Advisory Committee, 1939, p.14). In a few authorities women were employed in a welfare role, and the other aspects of management were distributed around other male-dominated departments.

Since the 1970s, however, two factors have had the potential to enhance the professional status of housing managers. These are the development of a more explicit and coherent view of the nature of housing work, and the improvement in the training and educational backgrounds of housing managers.

In the 1970s, as more and more housing departments were established in the new authorities created after local government re-organisation, their role became an issue in local government and in professional housing circles. Initially the idea of a 'comprehensive housing authority' was developed by, for example, the Seebohm and Cullingworth Committees (Committee on Local Authority and Allied Personal Social Services, 1968; Central Housing Advisory Committee, 1969). This idea emphasised the growing general role of housing authorities in improving housing conditions across all tenures, as well as through the management of the council's own stock. It was soon translated into administrative terms as the comprehensive housing department (Housing Services Advisory Group, 1978), but many authorities continued to allocate different aspects of housing work, especially repairs, to more than one department, and others divided the tasks between highly centralised sections within the housing department (Power, 1987, pp.84–8). The exact nature of housing work continued to perplex many of those affected by it or engaged in it. Advice from official inquiries such as those of Cullingworth (Central Housing Advisory Committee, 1969) with their clear prescription for an occupational role which encompassed social policy and economic aspects of housing as well as design and development issues, were ignored. That 'housing management' could constitute the coherent task of an occupational group with its own identity remained an unknown or unacceptable notion in many, especially smaller, authorities. Local authorities and central government obtained advice on housing management issues not from members of a recognised occupational group, but from civil servants and government advisory committees (Laffin, 1986). As recently as 1986 the Audit Commission recommended the appointment of a chief housing officer responsible for all housing management activities in authorities which still did not have such a post.

Throughout the period from 1919, qualified housing workers have never constituted more than a small proportion of people engaged in housing work. In 1986, of 55,000 housing staff in English local authorities, less than 2,500 were professionally qualified (Audit Commission, 1986, p.15). This is many more than for most of the postwar period, but it is no greater as a *proportion* than the 7.5 per cent found to have a professional qualification in 1977 (City University, 1977, p.8). Membership of the Chartered Institute of Housing comes closest to being seen as the accepted qualification, and some of our interviewees had acquired it at considerable personal cost:

'Five years correspondence course ... It was hard graft after doing a day's work, to come home, and many a day that's been a wet and cold day. The last thing on earth you felt like was sitting down with your books and – I have twin daughters, they're twenty-eight now, but they'd just been born. And that wasn't easy for the wife either, you know. She'd had them all day and I wasn't much of a help of a night time due to studying, you know.'

(Former Rent Collector, Dockport)

Some had perceived its value as an aid to promotion:

'When the vacancy arose I was told very politely but very firmly . . . "Well, you know if you'd been qualified, with your experience you'd have walked into that job". So I said to him "Right, then you just send me out to day release and I'll take the course and I'll be qualified and you won't be able to say that to me again." So I did.'

(Senior Officer, Braehead)

Another indicator of a stronger 'professional' orientation is that more housing managers have come into housing from higher education since the 1970s than previously – one of our citizenship authorities, Strathmuir, employed two graduates in its housing department by the mid-1970s, and now, in this large urban authority, graduates are almost too numerous to count. In addition, there has been an expansion in the provision of university level housing courses since the early 1980s, partly as a result of government support. And there appears to have been a growth in on-the-job training, with four out of five (81 per cent) staff reporting in one survey at least one day of training (5.6 days on average) in the last 12 months (Bines *et al.*, 1993, p.36). However, relatively larger numbers of graduates and qualified staff have been employed by housing associations than local authorities, possibly affecting the relative status of housing work in local government and housing associations (Bines *et al.*, 1993, p.35), and the majority of housing managers enter their occupation from school or some other job.

In the 1990s the precise nature of the housing manager's role is again a matter for controversy as the local housing authorities attempt to come to terms with the changes outlined in Chapter 1. For example, 'housing management' needed to be defined in order to specify the tasks to be carried out by housing management contractors under the CCT procedures. Other changes such as the development of a performance culture in local government, community care policies, the government's emphasis on the 'enabling' role, and the increasingly disadvantaged characteristics of council tenants, also raise questions about the extent to which housing managers are akin to social workers; the extent to which they are primarily concerned to collect rents, or ensure good property maintenance; and the extent to which they are co-ordinators of other housing providers. These changes led Cole and Furbey to conclude that:

> The radical restructuring and diminution of public housing of the Thatcher years occurred with little sign that the agenda had been modified by the representations of housing professionals.
>
> (1994, p.137)

This review shows that housing managers have a low professional status and an ambiguous role. Nevertheless, there have been attempts to sustain professional associations and to stimulate a stronger 'professional' ethic which values training and education. Attempts to define more clearly the nature of the tasks of housing management and therefore to define the professional skills and knowledge needed are, however, periodically overtaken by new developments which alter the environment in which housing managers work. This provides the necessary background to our more detailed discussion of the attitudes of housing managers, but it does not allow us to predict what the nature of the relationship between housing managers and tenants is likely to be. We now turn to this issue, using the data from our case studies.

HOUSING MANAGERS AND TENANTS

The crucial issue for many officers was the policy context within which they worked, which influenced their relationship with tenants. A typical comment of one officer, in a traditional authority, when asked the purpose of tenant consultation, was: 'Well the ultimate objective must be to make sure they (tenants) don't whiz off to somebody else' (Senior Officer, Talybont).

And an officer in another traditional authority perceived the motivation for officers' views as stemming from legislative and other changes:

> 'I don't think there's any resistance (to the idea of tenant consultation). Because I think unless they carry this out they are going to sink. It's as simple as that. I mean the pendulum has swung too far. Once upon a time we were ruling the roost, and we were kings. The pendulum swung (to) the other side. Unless we're prepared to be servants, if you like, I'm afraid

we're going to go under. That's basically what's going to happen.'

(Senior Officer, Dockport)

The specific context during our fieldwork was the enactment of Tenants' Choice. There is evidence that this has declined in importance in the minds of officers as tenants failed to take up the opportunities it offered. But other issues, such as the publication of performance indicators, and the introduction of CCT, have led to a continuing concern by officers about how the service they provide is seen by tenants:

many officers felt that there were growing expectations from their tenants, who it was felt were expecting higher levels of performance in service delivery than previously.

(Bines *et al.*, 1993, p.156)

More generally, housing managers were unwilling to see themselves as professionals who had to take no account of tenants' views. In expressing this view the language of consumerism was often apparent, with references to clients, customers, rights, and service. A typical comment came from a senior manager in a traditional authority:

'I am a professional housing person and I have a quite deep-rooted desire to offer a good service, and the only way you can do this is to . . . the only way the customer will think it is a good service is to give them what they want.'

(Housing Manager, Talybont)

One Director considered such matters relevant to the appointment of new housing management staff: 'I've always prided myself in a housing manager who has a commitment towards participation of clients in the delivery of a service' (Director, Compton).

Also common were comparisons with the private sector:

'I wouldn't agree (that professionals rather than tenants should manage housing). I mean, I would say that at the end of the day it's like anything – you gear your product to the consumer, and to all intents and purposes our product is the provision of suitable housing, and if the consumer – in this case the tenant – isn't happy where they're living and they give you complaints, then like any manufacturer, or what have you, you should investigate their complaints. If they're wrong you should tell them, and if they're legitimate you should try and do something about it.'

(Senior Officer, Dockport)

On the whole officers believed that this attitude was widely shared, or at least that any opposition to it was not expressed openly:

'I would say . . . that it's in our interest to talk to tenants, to talk to the people who pay our wages, and although we can't always grant their requests, we should have a realistic and regular conversation with them, a two-way conversation as well . . . But on the whole, I don't think (the opposite) attitude prevails really in housing . . . I may well have missed something, but it's not a view that I'm finding openly sort of discussed in housing circles.'

(Senior Officer, Talybont)

The view that housing managers or housing authorities would benefit from taking account of tenants' views was shared by officers at all levels of housing departments. But some qualifications were made about the reluctance of other actors – colleagues, councillors or tenants – to achieve a closer working relationship. For example, in one citizenship authority an estate officer saw resistance by some colleagues to any fuller involvement by tenants' groups in estate management:

'Some estate officers would automatically say that's a good idea, and they'd go along with it, and probably help them. Whereas others, who don't get on with the tenants' associations anyway, would probably be totally against it. Because you do get a clash of personalities, like anything (else).'

(Estate Officer, Compton)

In one of the traditional authorities, with very little experience of tenant participation, the expression of support for tenant participation was qualified by scepticism about tenants' interest in it:

'I suppose it would be good if there was more participation with tenants, but then it's got to come from the tenants too, that they want to be involved. I wouldn't say that anybody's really showing as such that they want groups set up, or we would – maybe the councillors more than ourselves – would have heard about it.'

(Assistant Housing Officer, Braehead)

63

Another type of qualification of their approval of a closer working relationship with tenants was the perception that in the short term it made officers' work more difficult. As one officer with responsibility for improvement and modernisation work said:

'At times, although they're quite within their rights and quite justified to give me those lists of complaints, they do tend to be a bit intolerant at times if they don't get a response right away, so they then write to the boss and say they haven't had a response. Well I may have only had the letter on the Monday, and because they haven't had a response by Friday there's this accusation that I'm just ignoring them, you see. So at the moment it definitely puts more pressure on me. At the end of the day, it should make the estate better, more manageable, and it will be a benefit.'

(Senior Officer, Dockport)

In contrast, the Director of one of the citizenship authorities, after recounting many examples of tenant participation in which he had been involved, said:

'Put succinctly, I think the point of tenant participation is to make sure we, the professionals have got it right. I think it balances our views and opinions and experiences by giving views, experiences and opinions of (tenants). I think without that we get imbalance, we get – all we get – is a highly professionalised service which could, and will, I think, show inadequacies in actually delivering what the community needs. I think they're very important, just to make sure we've got it right, and to help us to show where we've got it wrong.'

(Director, Compton)

So the overwhelming view of officers was that tenants had some role to play in housing management. The prevailing attitude was that if tenants wished to express a view they should be allowed to, and for many officers, taking account of tenants' views was a significant feature of what defined a 'professional' housing manager. These attitudes, with some variations, were found across all the case study authorities.

The reasons held for wanting to take account of tenants' views are varied. Some of the officers mentioned the use of tenant participation in helping towards the provision of a good housing management service, while others dwelt more specifically on problems such as difficult-to-let estates and neighbour disputes. Some officers saw participation as essential to the survival of council housing, while others saw it as leading to a change of attitude by tenants. Some officers combined more than one element in what they said:

> 'They (tenants) will take more interest in their homes and their surroundings, keep our property in a better state of repair. They should be happier in their homes and their environment.'
>
> (Senior Housing Officer, Dockport)

Tenant participation is pursued because it suits housing managers by making their life easier, or ensuring the continuance of council housing. Nevertheless, this is often tempered by a belief that tenants have legitimate rights as consumers of a service and at the end of the day housing managers realise they are managing the tenants' homes not their own. Therefore, the motivations of housing managers were a mixture of self-interest and altruism. However, the balance between the two elements varied considerably between different officers. Staff of traditional authorities tended to fall into two categories – those who knew little of tenant participation and found it difficult to express a view about it, and those who saw it as necessary for their own survival and to make their job easier. Consumerist and citizenship authorities had more staff who were prepared to view the involvement of tenants as a right, or who felt that tenants brought essential knowledge to assist in the difficult tasks of housing management. There was also a tendency for more senior, experienced and better educated staff to hold views associated with consumerism and citizenship – irrespective of where they worked. So officers did not necessarily hold the views which dominated in their authority. 'Consumerist' officers were found in 'traditional' authorities, and 'traditional' ones in 'citizenship' authorities.

WHAT SORT OF RELATIONSHIP?

As has been seen, most officers felt that tenants had some role to play in housing management. However, views on the way in which the relationship should be conducted sometimes differed greatly from one authority to another, and within authorities. But some similarities of approach could also be detected. These included a willingness to communicate with tenants' associations as well as individual tenants. However, even this apparent consensus could obscure differences of attitude and approach.

Many officers, including some in traditional authorities, attempted to combine communication with individual tenants and communication through tenants' associations.

'Individual letters is the best way (to inform tenants), I think, and that has been appreciated by the tenants, and then an individual tenant can come in and tell you his problem. Obviously if we've got a tenants' association in there we work with them and we discuss with them.'

(Chief Housing Officer, Dockport)

However, in the traditional authorities tenants' groups were not actively encouraged, perhaps reflecting their rather peripheral status in the relationship with tenants.

'In the ten years I've been here we've had the odd one (tenants' group) spring up for a particular purpose, and disappear again, and that's really how it's tended to be. We've certainly not initiated the setting up of any in the past.'

(Housing Officer, Talybont)

Similarly, in Braehead an area manager described how she had attended a meeting with some tenants who were concerned about conditions in their block of flats; afterwards she had one phone call from one tenant then she 'never heard of them again'.

In contrast, the two citizenship authorities both employed staff with a community work background and remit, to assist tenants to organise into associations whenever there was an expression of interest. In the circumstances of Braehead an officer in Strathmuir or Compton would have been asked to make contact with the tenants, to assist in setting up a tenants' association.

Officers in citizenship authorities stressed the importance of tenants' groups as a way of communicating with tenants generally:

'In the very beginning of this district (office) there were no tenants' groups at all, and I found myself being bothered on the same issues and having to go back to one hundred people, one thousand people, whereas if the groups are doing the job . . . they're representing the community. Together we try and resolve it.'

(Assistant District Manager, Strathmuir)

Some officers worried a good deal about the representation of tenants' associations and clearly felt this was a reason for not giving them much importance:

'You always find the tenants' associations that you get a few people come forward, it's not the entire – it's not representative for the whole area. Too many people just haven't got time and equipment to get stuck in and help.'

(Estate Officer, Compton)

Others were more resigned to working with them, in the absence of any more representative group:

'But, at the end of the day, it's maybe not for me to worry about how representative they are, because they're there in name, and there for people to use, and if other people don't want to use them – I mean, you couldn't have ten thousand tenants' associations. So, I mean, how representative is anything at the end of the day. So, you've mainly just got to work with who you've got.'

(Senior Housing Officer, Braehead)

Only a few officers, all in the two citizenship authorities, suggested ways of minimising unrepresentativeness. One officer, with a community development role, said:

'I actually believe that they're not unrepresentative – I actually believe they're not communicative. They don't have the ability to communicate very easily with their own, for want of a better word, constituency. I think the Council has a role to play there. I actually think that's an area where the Council ought to help them. And the way they ought to help them is very much in grants and funding and helping them with newsletters, and the production of them. And I hope very much that will be one of the things the Council take on board.'

(Estate Officer, Compton)

Officers' views about the forms the relationship with tenants should take appeared to be shaped by a number of factors including educational background, career experiences, and the dominant ethos of the authority. For example, the Directors of Strathmuir, Compton and Dockport had similar educational backgrounds and experience, but demonstrated very different attitudes to their relationship with tenants. In Strathmuir and Compton the Directors saw few limits to the form the relationship with tenants should take, emphasising the need for dialogue and negotiation before decisions were taken, and favouring joint management and management co-operatives amongst their models for the future management of council housing.

Their views about structures and methods for participation allowed for a similar variety, and they could refer to a range of liaison and joint management groups, at estate, area, and authority-wide levels, up to and including tenant participation in committees. This last form of representation was subsequently outlawed in England and Wales, but still remains possible in Scotland. In Strathmuir, though, the council had changed its view and removed tenants' representatives from two sub-committees in the early 1980s. In Compton, too, some disquiet was expressed, apparently out of sympathy with tenants' needs: 'It's almost tokenism having tenants there in a voting capacity, and it's intimidating for them' (Director, Compton).

The Director of Housing in the traditional authority of Dockport provided a marked contrast. He was also qualified and experienced in several authorities. When asked about the results of our household survey, which showed a large proportion of tenants who wanted more say in what the council did, he said:

> 'I don't really know what they're getting at ... it is difficult to find, because the majority of the council's accommodation is in what I call the Labour political stronghold of the Borough, and the Labour councillors are very active, knowing their community politics, and they do bring forward the issues that are going on ... so I think it's information tenants want, and yet there's never been a strong call in Dockport, people have not been looking for information in any great degree ...'

and when asked about involving tenants in estate management structures, such as Estate Management Boards: 'Well I suppose I am saying I don't believe in it, I don't believe that's what the tenants want, it's never surfaced ...' (Director, Dockport).

This officer acted on this belief by actively limiting tenant participation to fit his views. Therefore, he refused to speak to a tenants' federation because of their wish to be consulted on policy issues. As we shall see in Chapter 6 his attitude towards tenants' groups was heavily criticised by tenant activists. The main differences between the Dockport Director and the two others were that he had worked in smaller authorities than they, previous to their current posts, and he now worked in one dominated by his own and the councillors' 'traditional' attitudes in contrast with the 'citizenship' attitudes common in Strathmuir and Compton.

There were also significant differences between officers in relation to the issues that should be discussed with tenants. Probably the most popular with officers was capital improvements, although opinions varied as to whether this should involve telling tenants what was to happen to them, consulting them, or involving them in developing proposals and plans. Many officers had reservations about discussing with tenants issues seen as sensitive, such as allocations, although some saw ways in which this could be done, as in Strathmuir which had experience of local lettings policies involving tenants. Officers here and in Compton – the two citizenship authorities – seemed inclined to go furthest in considering tenant involvement in aspects of allocation policy making and administration, but even here officers had reservations. The ambivalence was expressed by one:

> 'Again, how far are you going to take that? I don't know – it all comes down to education, doesn't it, and what you're expected to do as an authority? And you've got to let them (tenants) be aware of the obligations under the Race Relations Act, and whatever else. And, I suppose, if you get people on it that are sensible, okay. But you don't, you get a lot of people sometimes that are a bit stubborn and set in their ways, so I don't know how far you would take that. I don't know, it would be difficult. I'd certainly – as I say, if it was done properly, you know, I can't see any problem with it, I mean, (the nearby tenant management co-operative) obviously works.'
>
> (Estate Officer, Compton)

The contrast in views about the issues tenants should be consulted or involved in can be illustrated with reference to the issue of the colour of paint:

> 'I have a feeling that the tenants' associations will start saying they want to pick the colour scheme. Now my reaction to that is well, okay, if they start asking we'll look into it, but strictly speaking, if we've got somebody who's capable of making a decision on painting that's in accordance with the surround-ings, and what have you, and it does look good, I would tend to say that something like that should just be ordered and we don't have to start writing off to them.'
>
> (Housing Officer, Dockport)

In another traditional authority – Braehead – the same attitude had prevailed until a newly appointed Senior Housing Officer – with atypical views for his authority – had been based in an office in the largest and least attractive housing estate in the area. He said:

'I think they (tenants) should (have an influence), yes. I think you've only got to look at the balconies down in King Street, and the colour down there. And no tenant was consulted about what colour the balconies should be. And they're now this kind of yellow colour. Well, the balconies up here, for example, people were given a limited choice between three or four colours. And as a consequence, well, I think, they got a sensible colour, and people are happy with it. Whereas in King Street people grumble about the yellow balcony.'

(Senior Housing Officer, Braehead)

So what appeared earlier to be an overwhelming consensus about tenant participation can be seen to break down in the more detailed elaborations of what that means. The main differences were about the approach to tenants' associations, the housing topics which should be the subject of tenant involvement, and the extent of the opportunity to influence policy which tenants should be given. This last difference translated on the ground to an approach in the traditional authorities, for example, which stopped short of any effort to develop a dialogue except perhaps over capital improvements, and which did not favour the setting up of tenant management organisations, unless tenants demanded it (which they never did). In the citizenship authorities, in contrast, contact between tenants' associations and housing managers was more frequent, many more groups existed, and new forms of participation involving tenants serving on committees, and in other ways, had been encouraged and fostered.

However, despite the differences, some officers in all types of authority felt that their own job satisfaction required account to be taken of tenants' views. For example:

'Wouldn't it be nice to know that if you'd done a ballot of these people, and said "Are you happy with the service you get? Are you happy living where you are?" and they all said "We're very happy. We think the landlord is great. We wouldn't swap him for anything." I mean, there is a great deal of satisfaction in that. Because you know that you are not wasting your time.'

(Housing Officer, Dockport)

However, job satisfaction as a result of tenant participation was seen more as a fortuitous outcome, rather than the primary reason for involving tenants.

TRIPARTITE RELATIONSHIPS

There is a large literature on the relationship between officers and elected members in local government, concerned largely with the extent to which one or other is the more influential, or the ways in which alliances, such as those between senior officers and elected members, may influence the policy process. This literature tends to neglect the role of consumers in influencing the relationship between officers and councillors. On the evidence of our interviews, the relationship between councillors and housing managers cannot be understood without reference to tenants. One officer, when asked about the tenants' role in housing management said:

> 'I don't think tenant participation is taking over the professionals' job, I think tenant participation is to have a third input, if you like, and that input to be considered equally with all other inputs ... I've never had the viewpoint that we just followed a line without taking into account the tenants' views. I've never done that – got myself into bother occasionally if I've taken the tenants' views.'
>
> (Senior Housing Officer, Braehead)

The perception of a tripartite relationship was important to many officers who were very aware that they did not provide the only form of access to the council, even in traditional authorities where there was evidence mainly of councillors raising issues as a result of being approached in their traditional role as elected representatives, rather than as a result of seeking tenants' views actively.

The nature of the tripartite relationship was illuminated when officers were asked about how they handled requests or complaints from tenants. Broadly, if officers – particularly senior ones – agreed with the tenants' view they would try to act on it, if necessary by seeking to change policy. But if they did not agree, they were conscious that tenants could go to councillors instead. One Deputy Director described how he would handle a request from a tenants' group:

> 'The senior officers all meet once a month – and these matters would come up through that source. We would discuss them and if we felt, like ... it was not something that would be acceptable ... we would say as officers "Now in our estimation this is not going to be workable because ... " and we'd explain why "but nevertheless, if you don't accept that and you feel so

strongly about it, we're quite willing to put it to the Housing Committee and ask their view" because we are not the ultimate (authority).'

(Deputy Director, Braehead)

Another said that if officers did not support the tenants' view he would tell them how to take the issue further:

'"You do know you've got the Liaison Committee, you've got Ward members and you've got other elected members. If you want to take it up, do so, but if it does end up at a Committee, you won't get our officers' backing to it, but again that's your (affair)." Just let them know where they stand . . . More than that though, I think I'd like to let them know the effects of what they're asking rather than say "I don't think it's a good idea". I would prefer to go to some lengths to explain why we didn't think it was a good idea.'

(Housing Officer, Sainsborough)

However, elsewhere in the same consumerist council, an estate officer indicated he would leave it to tenants' groups to take things further if they wished. He said tenants' groups should approach his council through the District Manager, but if they did not achieve what they wanted they would: 'probably go to the councillors then, . . . or to the Director. That's the normal routine I've seen over the years' (Estate Officer, Sainsborough).

Officers in the citizenship authorities were most used to finding tenants taking an alternative route to the council through councillors, and they did not always like it. A senior officer in Strathmuir said he was accountable to: 'the Housing (department) Directorate, my local (elected) members . . . (and) I am totally accountable to our tenants.' He acknowledged this could sometimes lead to dilemmas and difficulties, such as when tenants and officers could not reach agreement on an issue, and the tenants 'insist, and then they go to the elected member, and against advice from officers it goes through, and the end result is disaster' . . . 'I cannot quote you an instance where a councillor did not respond to pressure from tenants' (Senior Housing Officer, Strathmuir).

Officers were clearly influenced by the knowledge that tenants may appeal to councillors if they are not satisfied with the treatment received from officers. Even if tenants did not actually go to councillors the fact they might have done seemed to make a difference to officers. In the traditional authorities this is most likely to be in relation to routine matters of housing management affecting individual tenants. In the citizenship authorities it might also be in connection with issues raised by tenants' associations. This raises a number of questions about the motivations of tenants, and about the role of councillors.

VIEWS OF TENANTS' MOTIVATIONS

Officers had more simplistic views about tenants' motivations in the council–tenant relationship than tenants themselves expressed (see Chapter 5). Overwhelmingly officers perceived tenants to be motivated by a desire to achieve something for themselves and their own local neighbourhood rather than improving the lot of tenants in general.

'I think the incentive is that you, in theory, are going to be living in that area, possibly for the rest of your days . . . You have to have children there, you have to bring them up in that environment. I think that it's only right that you should have some say.'

(Housing Officer, Dockport)

This was sometimes expressed in pejorative tones, with some officers expecting a greater degree of altruism in seeking to improve other tenants' housing conditions than some tenants displayed. One of the classical arguments in favour of democratic participation – to defend one's own interests directly – is not acknowledged by some housing managers. Other officers were concerned about the perceived style of some tenant activists: 'Some are genuinely very concerned for their area and wish to be involved in the process. Others are power crazy' (Estate Officer, Strathmuir).

'As a younger housing manager I'd have said . . . there are people who sort of want to come in to boost their egos, to sort of feel good, etc., but now I'm not sure because I look at the people I know well within the tenants' movement (here) and I'd venture to suggest that they represent all types.'

(Director, Compton)

Several officers thought they detected changes in people as a result of their closer involvement. One senior officer, in a traditional authority, who had limited experience, perceived differences in the attitudes of tenants over time, as they were involved in discussions:

'When you first go and talk to tenants they're quite full of animosity because they want this, they want that, "It doesn't matter to me what you're doing over there, it's this place we're talking about, what are you going to do about it?" After a couple of times, three times, tenants' viewpoints begin to change and they look on the broader aspect of the thing because they begin to understand. The officer then begins to understand a lot better as well. The two then begin to work together to get what they can within the parameters that are set for them.'

(Senior Housing Officer, Braehead)

An officer in Dockport also detected a change in tenants: 'They realise that you're only adopting the policy, whereas before I used to get a lot of letters that were directed personally at me.' He went on to say that he had changed as well:

'I think, from a personal point of view, I think I'm more ready to recognise the problems they're having, whereas before I just thought "Well, I've just got to look after (the Council)" . . . I tend to give more credence, if you like, to some of the things they're putting forward.'

(Housing Officer, Dockport)

Although experience of tenant participation did not always lead to a more tolerant view of the motivations of tenants, in some cases it did. This arose in two different ways. First, officers were able to have more empathy with tenants as they got to know them better. And, second, officers detected a learning process at work, in which tenants were finding out how to be more effective, or were coming to appreciate the problems of the council, or the problems of other tenants. However, there was a perception by some officers that attitude change in the longer term depended on tenant participation producing results on the ground:

'Good participation is essentially participation which achieves results on the ground. It can result in attitudinal change, but it is practical results which bring about this change. Perhaps if they (tenants) were anti-local authority before and all that's happened is that they have had a chat about it with the housing manager, nothing has actually happened but they have had a nice time, had a cup of tea and a biscuit, then it won't fundamentally change their attitude. What would change it would be more trees on their estate a year after they asked for them!'

(Housing Officer, Dockport)

The impact of the relationship between tenants, councillors and housing managers on the parties' attitudes towards each other is explored further in Chapter 7.

CONCLUSIONS

This chapter has shown that housing managers do not conform to the traditional view of professionalism, in that they acknowledge a role for tenants in their work. The language of private sector customer service has had an important influence on them, and perhaps their attitudes, while legislation and other changes in housing policy have influenced housing managers' perceptions about the need to provide a service which satisfies tenants' aspirations. However, differences exist in housing managers' approaches to taking account of tenants' views. At one extreme, an apparently reluctant Director is pushed into consultation over modernisation projects, and resists more far-reaching forms of tenant involvement; at the other, two Directors commit staff and other resources to developing a wide range of methods and structures for participation, with scope for tenant involvement over issues such as rents, allocations and repairs as well as modernisation, and see few limits to the future development of tenant involvement. Both citizenship authorities employed staff with community work backgrounds, and none of the other authorities did. Officers in traditional authorities were more likely to perceive tenant apathy, and cite it as a reason for the slow development of tenant participation. Differences of attitude to tenants' groups also existed, with those who feared their unrepresentativeness at one end of a spectrum of views, and those who wished to provide resources to support their development at the other. There was, though, wide acknowledgement by officers in all authorities of a possible role for tenants' associations, sometimes because housing managers saw ways in which they could benefit themselves.

The motivations of housing managers for involving tenants were a mixture of self-interest and a recognition of the knowledge and rights of tenants. Tenant participation was widely seen to lead to a better housing management service and in the long term to make housing managers' jobs easier by reducing management problems on estates by providing a means of communication to managers as well as changing the attitudes of tenants. At the same time, enlisting the support of tenants was seen by many as crucial to the survival of council housing and, therefore, to the continuance of their own jobs. Nevertheless, many housing managers thought that tenants had a right to be involved in housing management because it affected the tenants' homes. However, views on the nature of this right and its extent varied considerably. The balance in the motivations of housing managers between self-interest and altruism was different for individual housing managers and there were variations between different types of authority. In traditional authorities there was less emphasis on the rights of tenants than in the other kinds of authorities.

This chapter has shown that the nature of the relationship between officers and tenants cannot be considered without reference to councillors who provide 'a third input', or possibly an alternative channel of communication for tenants. The accountability of housing managers to tenants in council housing seems to be enhanced by the possibility that tenants will appeal to councillors if they are not satisfied with the housing manager. This raises important issues about the accountability of housing managers working for other types of landlord to which we return in the final chapter.

In summary, the role seen by housing managers can be best characterised as one in which they are influenced by both councillors and tenants as well as their own professional ethics. More senior officers, and better educated officers are more influenced by current policy developments and thinking about the role of consumers than junior, less well trained officers. But some officers perceive a bigger role for tenants than others, with reference to the forms, structures and methods of participation, the housing issues to be discussed, and the role seen for tenants' associations. The more experienced in tenant involvement the officer the more likely it is that he or she supports it. But perhaps the most important factor is the ethos within an authority, and crucial to that is the attitudes of councillors. It is to these we now turn.

4

COUNCILLORS

Councillors have a formal authority as elected representatives in British local government which neither professional expertise nor citizen involvement can usurp. Yet tenant participation – and other forms of citizen involvement – provides just such a challenge; and councillors themselves had often initiated the new relationships developed with tenants in our case study consumerist and citizenship authorities. This chapter starts to explain this apparent paradox by considering how councillors see their role and the implications this has for their relationship with tenants and housing managers. We later return to the nature of the relationships involved in tenant participation in Part III, where an analytical framework for power is used to consider the puzzle of how councillors' relationships with tenants and housing managers can deviate from the traditional representative model.

The chapter starts with a review of who councillors are and what they do, in the context of a highly politicised local government system in most parts of Britain. It goes on to outline the typologies which have been developed by political scientists to assist understanding of how councillors go about their work and view their role, and proposes a refinement to take account of consumerist and citizenship attitudes. The perceptions of tenant participation held by councillors in the six local authorities are then discussed and we also examine councillors' views of housing managers' attitudes. Finally, conclusions are drawn about the differences in the ways councillors see their role in relation to tenants.

Other studies of local politics by political scientists have also examined how councillors act and perceive their role (Hampton, 1970; Dearlove, 1973; Newton, 1976). And recent studies of councillors have built on these writings (Barron, Crawley and Wood, 1991;

Bloch, 1992). This has led to the development of a number of typologies which are intended to assist in conceptualising the work, motivations and attitudes of councillors (Newton, 1976; Gyford, 1984).

WHO ARE THE COUNCILLORS?

There appears to be a tension in the traditional view of the role of the councillor – as someone who could be approached or used as a channel for the conduit of grievances, feedback or comment – and the ideas about tenants' rights, consultation, involvement and consumer control which characterise the current developments in local government outlined in Chapter 1. In addition, if tenant participation is seen as a way of altering the relationship between councillors and tenants towards giving tenants more influence and power, there is plenty of scope for councillors to feel threatened or defensive. A traditional councillor would not favour much direct input by tenants into housing management, but may see his or her role as being to feed tenants' views into decision-making when he or she considers appropriate. But a consumerist councillor, in contrast, would favour the creation of mechanisms whereby individual tenants but not tenants' groups could make their views known. A citizenship councillor would support arrangements for individual tenants and tenants' groups to engage in dialogue and negotiation.

Councillors have been the focus of a good deal of attention by social scientists assisted by three major postal surveys carried out each decade since 1964. These surveys were used to assist government inquiries into aspects of local government chaired by Maud (1967), Robinson (1977), and Widdicombe (1986), and have provided data for analysis in academic writings (for example, Gyford, Leach and Game, 1989; Young and Davies, 1990).

Councillors are said to be 'overwhelmingly male, middle aged, middle class – and, of course, white' (Gyford, Leach and Game, 1989, p.45). The background and socio-economic characteristics of councillors have changed little since the 1960s. By 1986 one in five (19 per cent) councillors was a woman; over one half (52 per cent) were aged between 45 and 64 compared with 29 per cent of the population; only one in five (21 per cent) of those in employment occupied a manual job, compared with 47 per cent of the population; and almost one in three (31 per cent) had been educated to higher education levels, compared with 13 per cent of the population

COUNCILLORS

(Widdicombe, 1986, Chapter 2). Councillors are more likely to be owner-occupiers and less likely to be council tenants than the general population. Young people are under-represented – only one in four (26 per cent) were aged up to 44, compared with 46 per cent of the population. Ethnic origin was not included in the Widdicombe or previous surveys, but Gyford *et al.* estimate that there were no more than 250 Afro-Caribbean or Asian councillors (no more than one per cent of the total) in 1986, compared with 4.2 per cent in the population as a whole (Gyford *et al.*, 1989, p.48). 'Despite the small changes, the most remarkable aspect is the stability of the population (of councillors) . . . elected members as a group are still highly unrepresentative of the overall population' (Widdicombe, 1986, p.19).

Few councillors have current experience of being a council tenant, although some may have lived in a council house in the past. If representativeness is seen as being about reflecting the socioeconomic characteristics of the electorate then the traditional view of the councillor's role as representative may be challenged, and often is as we shall see in the next two chapters. On the other hand, differences between elected representatives and the electorate have existed since elections became commonplace, and this has arguably had relatively little impact on the legitimacy of elections.

PARTY POLITICS

The personal characteristics of councillors represent only part of the context within which tenant–landlord relationships develop in local government. At least as important is the party political nature of local government. Although party politics is 'all but universal' in British local government, outside the most rural areas (Young and Davies, 1990, p.27), the impact varied from one locality to another. In their 1989 survey Young and Davies found that the reality of party politics is increasingly recognised in the formal working and management arrangements of local government throughout Britain, but the pattern is one of a 'sharpening of party differences in the more partisan urban areas, coupled with the persistence of independent patterns of behaviour in the more rural districts . . . Evidently what is happening in local government today is a greater differentiation of political style, with the long term trend towards universal political competition masking a growing intensity of politics in urban areas, and by and large *only* in urban areas' (emphasis in original) (Young and Davies, 1990, p.51).

79

This was reflected in our case study areas which included four highly urban, politicised authorities, and two with a more rural hinterland, but a strong tradition of party politics in local government. The authorities varied, however, in the extent to which they were polarised by the extension of party organisation into all facets of the council's work, for example, by party members mandating councillors about how to vote, or by party groups deciding on issues before sub-committee meetings as well as full council meetings. The most polarised – most obviously controlled by party groups opposed by other party groups – were Strathmuir, Compton and Sainsborough, the citizenship and consumerist authorities. Housing was seen as a particularly contentious issue, in a party political sense, in these authorities. 'It has been the one (committee) that's been most bitterly divided between the groups and the least willing to compromise' (Councillor, Sainsborough). This reflects the extent to which the housing role of local government has been at the forefront in debates about the role of the state since 1979.

All the councillors we interviewed were elected as members of a political party, although one had reluctantly declared his Conservative Party allegiance in order to avoid being opposed by his own party, and another had not originally stood, 27 years previously, as a Liberal Party candidate, although he had been a member of the party all that time. Politics played an important part in motivating many to serve in local government in the first place, as well as in influencing their behaviour, and the management of the authorities on which they served. This did not prevent occasional deviations from the party line in some cases:

'I see myself as more loyal to the community than the Party ... I may be a member of the Party but it's the tenants that voted for me. And I have broken the whip on issues that I think were important to the tenants ... '
(Councillor, Compton)

This councillor had been a member of her political party for two decades, since she was in her teens, as had another, who saw the move onto the council as 'a natural progression' (Councillor, Sainsborough). But not many would have seen the transition to councillor as quite as smooth and inevitable as one did, in a highly polarised and party politicised urban authority: 'It's a very thin line that divides you from being a councillor, if you are involved fully in the Labour Party anyway' (Councillor, Compton).

All major political parties support tenant participation, with some differences in the nature of that support, but councillors cannot be assumed to know their party's policy, let alone support it in any detail. Party policy on housing generally is likely to be more influential than policies of individual local authorities on participation. Nevertheless, councillors' attitudes to tenants and to tenant participation may reflect views of the role of council housing which themselves may be related to general political ideologies. We showed in Chapter 2 that many of the ideas behind consumerism and citizenship as applied in local government have their roots in general political philosophies.

WHAT DO THEY DO?

Councillors spend many hours on council business. When asked (for the Widdicombe Committee) about the numbers of hours spent in a typical month, the average was 74 hours, roughly equivalent to a half time job. This average masked wide variations in the average time spent by councillors on different types of authorities, from 129 hours per month spent in Scottish Regional and Islands Councils to 58 hours spent by members of English shire districts (Widdicombe, 1986, pp. 42–53).

The general picture is that councillors in Scotland and Wales spend more time than councillors in England; councillors on upper tier authorities spend longer than those on lower tier (housing) authorities; English metropolitan authority councillors spend longer than English shire district councillors. In relation to party background, Labour councillors spend longer than Conservatives and Liberals. The older councillors spend longer than the younger ones; the unemployed spend longer than those in employment; and office holders (such as committee chairs) spend longer than backbenchers. The time spent is disputed by Barron et al. (1991) who found that the time recorded in the diaries by county councillors was on average 137 hours per month, while the time spent which could be recalled was 82 hours per month.

How much of councillors' time is spent on different activities was reported by Widdicombe under seven headings. Over half (39 hours per month) was spent at council meetings or preparing for them. Dealing with electors' problems, surgeries and pressure groups took 13 hours per month. Meeting external organisations occupied eight hours; and public consultation meetings occupied two hours, on average (Widdicombe, 1986, p.42). This categorisation is not very helpful in clarifying the time spent on participation activities.

The evidence of the Widdicombe and Robinson Committees is that councillors spent slightly less time on council business in the mid-1980s than ten years previously. Other evidence (reviewed by Gyford *et al.*, 1989, pp. 61–8) suggests that is not the case. Further, they argue that councillors feel increasing pressure in the conduct of their work which is reflected in increasing voluntary resignations. These pressures are a result of frustration at the party politicisation of local government which was particularly felt by older Conservative and independent members; frustration about exclusion from the leading group of the majority party, felt by members of majority and minority parties; the perceived dominance of officers; and disillusion at central government restraint and interference, felt by Conservative as well as other councillors, and leading to a reluctance to stand or stand again for the council. These feelings were echoed by many of the councillors interviewed in our six areas.

Nowhere in the literature is there any suggestion that councillors are frustrated or disillusioned by the development of consumerism or public participation, although there is evidence that some councillors feel that public or interest group involvement or consultation has gone far enough. The Widdicombe survey found that 31 per cent of councillors felt that wider co-option would be beneficial, but 50 per cent felt it would not. Roughly the same number agreed (40 per cent) as disagreed (42 per cent) that too little consultation was held with interested parties before decisions were taken. And more (57 per cent) agreed than disagreed (21 per cent) that ordinary citizens should have more say in the decisions made by local government. Co-option was favoured by Liberal and Labour more than by Conservative councillors, by 'backbenchers' more than holders of offices, and by less experienced councillors more than long serving ones. Consultation with 'interested parties' was also favoured more by Labour, and especially, Liberal councillors; by minority party councillors more than majority party councillors, and by less experienced councillors more than long serving ones (Widdicombe, 1986, pp. 72–5).

TYPOLOGIES

These and other findings have led political scientists to develop typologies which describe the variations in attitudes to their role displayed by councillors. The most common division drawn is between 'tribunes' and 'statespeople'. 'Tribunes' concentrate on their representative role in relation to their ward, while 'statespeople' are primarily interested in overall policy making and implementation. 'Tribunes' are typically less experienced; 'statespeople' often hold formal positions either within the political party group or chairing a committee or sub-committee, for example.

Gyford (1984) has elaborated this typology with reference to no fewer than thirteen aspects of the councillor's role and work. For example, 'tribune' councillors are said to be likely to display a 'watchdog' attitude in their relations with officers, and 'statespeople' councillors are expected to treat officers as colleagues. More relevant to our concerns are the councillors' attitudes to relations with constituents, and relations with interest groups. The 'tribune' is expected to be a 'welfare officer' for constituents, and a 'facilitator and spokesperson' for interest groups. The 'statesman' is expected to be a 'mentor' with constituents, and a 'resistor and referee' with interest groups (Gyford, 1984, p.18).

These are intriguing and complex conceptions, especially in relation to interest groups. The 'facilitator' and 'spokesperson' roles are not the same, despite Gyford's attempt to bracket them together. 'Facilitator' suggests assisting an interest group to present a case, whereas 'spokesperson' suggests making it for them. Equally 'resistor' and 'referee' imply, on the one hand, resistance to an interest group's approaches and, on the other, a neutral stance and an arbitration role between opposing parties, which might involve facilitating the interest group's approach to the authority. 'Facilitators' and 'referees' are therefore likely to be well disposed to ideas about consumerism and citizenship, whereas 'spokespeople' and 'resistors' imply traditional notions of the representative function. Gyford's typology hints at, but does not elaborate on, the possibility of an attitude towards interest groups which would facilitate and assist communication with the authority, and which 'tribunes' and 'statespeople' might, in theory at least, share.

Newton (1976) identifies six different dimensions of the councillor's role:

1 attitudes towards the nature of representation
2 behaviour within council groups
3 preference for governing the city as a whole or looking after the interests of a ward

4 preference for dealing with general policy issues or with individual problems

5 preference for specialising in one aspect of council work or generalising over them all and

6 attitudes towards the involvement of community organisations in the government of the city.

Numbers 1, 2 and 6 offer the best prospect for examining attitudes to participation. Newton provided a fivefold typology of councillors based on the attitudes of Birmingham councillors to the six dimensions above. These are

- parochial
- people's agent
- policy advocate
- policy broker
- policy 'spokesman' (sic).

The parochial, the people's agent and the policy advocate all suggest a positive attitude to raising in the council at least some of the issues brought to the attention of the councillor. But these represent three different aspects of the traditional representative role, rather than necessarily a positive attitude to consumerism or citizenship, which involve tenants making their own views heard. The policy brokers are 'more likely to perform the classical political role "of mediator and reconciler of different interests"' (Hampton, 1991, p.125). This is the nearest Newton gets to examining the idea of councillors as facilitators, whether at ward or general policy level, and the idea of councillors as communicators from council to people. Whether this neglect reflects a lack of empirical evidence to support a typology based on the role of councillors in a citizenship or consumerist authority is not clear. It may be that Newton's research, carried out in the 1970s, could not have detected the later development of role orientations concerned with citizenship and consumerism. Our task was to examine what role orientations could be detected in our case studies.

The existing typologies do not seem to take adequate account of the possibility of councillors having a role in facilitating the participation of citizens *irrespective* of what the councillors think of the groups/individuals, their views, or how they have expressed them. The typologies also tend to present an over-simplistic model of decision-making processes in which all a citizen or interest group has to do is persuade a councillor to support their case and all will be well. In practice a multiplicity of councillors, officers and others are involved in developing policy, or handling individual complaints, and councillors are at best normally able to influence events as one of a number of actors. Councillors could have a role in introducing a new set of actors – the tenants – in a variety of ways which the typologies of councillors do not illuminate. A new typology can thus be developed, showing councillors' possible role orientations in relation to tenant participation, and distinguishing their attitude to relations with tenants from their attitude to relations with tenants' associations.

As Table 4.1 shows, there are less important distinctions to be drawn between tribune and statesperson councillors than between traditional, consumerist and citizenship ideal types in relation to tenant participation. Ideal typical citizenship councillors would demonstrate a willingness to facilitate or referee a process in which individual tenants or tenants' representatives were assisted to gain access to the authority. This might, for example, involve a councillor assisting a tenant to find the right officer to discuss her tenancy problem with, or convening a meeting where officers and tenants' association representatives seek to thrash out an acceptable procedure for handling the removal of asbestos, as occurred in Strathmuir. In contrast, traditional councillors would either assist tenants with advice, or would decide whether to agree to speak on behalf of tenants' representatives in the council. The consumerist councillor would share the same attitudes as the citizenship councillors towards individual tenants, but in relation to tenants' associations would act as traditional councillors.

This reformulation of part of Gyford's typology is an ideal type and not, therefore, 'true' or 'untrue'. The extent to which councillors

Table 4.1 Typology of councillor role orientations to tenant participation

	Relations with tenants	Relations with tenants' associations
(a) Traditional tribune	Welfare officer	Spokesperson
(b) Traditional statesperson	Mentor	Resistor
(c) Consumerist tribune	Facilitator	Spokesperson
(d) Consumerist statesperson	Facilitator	Resistor
(e) Citizenship tribune	Facilitator	Community worker
(f) Citizenship statesperson	Facilitator	Referee

in general might be categorised into each cell of the typology is unknown. Our interviews are insufficient in number to justify any but a tentative conclusion that at least some councillors appear to adopt each of the role orientations specified in the typology, as we now demonstrate.

The citizenship councillor would be likely to have a commitment to 'facilitating' *and* 'refereeing', which Gyford places at opposite ends of a continuum concerned with attitudes towards interest groups. The consumerist councillor would assist consumers in the exercise of their rights and choices, rather than act on their behalf, as in the paternalistic, welfare officer role of Gyford's tribune councillors. We now examine whether these attitudes were apparent in the interviews we conducted.

ATTITUDES TO THE ROLE OF COUNCILLOR AND TO TENANTS

Some councillors in our study found it hard to decide whether they were predominantly a 'tribune' or a 'statesperson' councillor, not surprisingly: 'I think it's fifty-fifty' (Councillor, Strathmuir). Some were anxious to acknowledge the dual role, even when one role dominated. For example, one minority party group leader who acknowledged that 'policy has to take up more (time)' also said:

'One does one's best for constituency problems, obviously. It's a high priority with our whole group because none of us could have got in without community campaigning and taking up the casework problems.'

(Councillor, Sainsborough)

Some councillors found it easy to say that their ward work came first: 'I'm interested in the case work side of being a councillor. I think that is the most important side of being a councillor' (Councillor, Braehead). Others said the policy element was equally important, or that the two were inseparable, while only a few said unequivocally that policy work came first. But whether they were 'tribune' or 'statesperson', councillors all felt it necessary to make efforts to keep in touch with their constituents' concerns, whether by holding regular surgeries or in other ways. For 'statesperson' councillors the constituency extended to the boundaries of the authority.

The role of the councillor in relation to tenants was described by a councillor in a citizenship authority:

86

'I would basically talk to the officers, having found out what the actual tenants wanted, their views, and then relayed them to the officers . . . If I felt really strongly the tenants had a case I would then go back to housing (a second time) and say "Well, all right, I'm not happy, I'll take it further. I'd like to go to Housing Committee. And I give you notice that I will submit a petition on behalf of such and such an estate 'cos I think they're right and I think we're wrong."'

(Councillor, Compton)

This classical statement of the traditional role of the councillor was replicated in different forms in all the case study authorities. But some councillors argued that tenants should have a voice of their own, and should not have to depend solely on the councillor to convey complaints or comments to the authority. For example, one Labour councillor in traditional Talybont saw a potential clash between the traditional, representative, role and an alternative approach with direct tenant participation:

'Well, the current policy is, of course, that there's no direct involvement of tenants – and that's to be regretted, I think – other than through their elected representatives. I've got to say some councillors think that's the way it should be. And there's a lot to be said for that. I am the person responsible for council house tenants, if you like. I am elected by them. They are in my ward. They've got a problem – I represent their views through the Housing Committee and through the Sub-committee, and so on. And yes, that in a sense is good . . . But I think we could involve (tenants) a lot more. Sadly, one of the proposals that (a senior housing officer) put forward a year or two back about area committees wasn't enthusiastically taken on board by some of the (majority) Labour Group, some of the councillors. They saw themselves losing a little bit of power, I think, you know.'

(Councillor, Talybont)

Another councillor, a Liberal Democrat, in a hung council, consumerist Sainsborough, insisted that tenants should be able to have their say even if he disagreed with their point of view:

'I think it's important that what they say is taken into account and that maybe I've been wrong in disagreeing with them and they've been right and it would be very unfair if I acted as a

87

censor and said "No, your views are not going to be heard by me" . . . If I can quote a non-housing example, I actually got a man planning permission for something I opposed. I told him how to go about lobbying to get what he wanted and he was very successful.'

(Councillor, Sainsborough)

This councillor demonstrated attitudes which clearly characterise the consumerist council, and he also went on to describe similar attitudes to facilitating tenants' groups' access to the council, which suggests a citizenship approach to facilitating and communication. This attitude, akin to the philosophy of community development work, was demonstrated also in Strathmuir and Compton, where a councillor described a working party with tenants' association, officer and councillor membership:

'Now it was obviously, I think, important to get the support of the tenants to have natural ventilation put in, and that working party was one way of doing that, and on that working party, I mean, I would be largely guided by what the tenants actually wanted. I mean, I could have gone along to a meeting with a preset view that I don't think it's appropriate to have natural ventilation, but if the clear majority of tenants on that working party said "We think its a good idea that we try natural ventilation" then that would be fine by me and I would be guided by what they wanted to do, and I would then see my role as being *to make sure that the council officers were clear that that's what the view of the tenants was.* So I think one of the things you find is that the ability that some council tenants have to express themselves is not as good as – as my ability to express myself is.' (our emphasis)

(Councillor, Compton)

This role of facilitator, however, can work in more ways than one. This councillor went on:

'But also, you know, it may have to work the other way, that if council officers say, you know, the lift shafts are crumbling and work needs to be done, and that means the lifts have to be out for three months, then it's, you know, the tenants may say "We don't want the lifts to be out of action" and then you have to work it the other way and say "Well, you know, you've got to be quite clear, if the lifts are not repaired, you know,

you may be travelling in it one day and you may come crashing
to the bottom of it." So it does have to work the other way,
I think, and that's how I see our role in that.'

(Councillor, Compton)

This view of the councillor as facilitator or community worker was
not commonplace anywhere. It was held more frequently in the citizenship authorities than elsewhere, but could also be found in the
traditional authorities:

'I sometimes argue with other elected members that we
shouldn't take people along by the hand and do it (helping)
for them. The reason I would argue that is because then the
next time it (a problem) comes up, they're in the same position
and this time they may not have the help from officials or
from ourselves. The best way is for people being able to do it
for themselves, or to be able to have the education, training,
the information, the resources, to do it for themselves.'

(Councillor, Braehead)

Councillors like this one detected differences in attitudes which
some attributed to party political differences, and others to the dominance of a professional ethos. For example:

'The previous Conservative regime's attitude was "I will speak
to an individual tenant but I'm not having an organisation talk
to me" ... I don't think it was helpful to anybody ...'
(Communication with tenants' groups) 'means that you have
the opportunity for an interchange of ideas and information
and understanding.'

(Councillor, Sainsborough)

'A trade union represents the rights of an employee, and a
tenants' association represents the right of the tenant to the
landlord, and the Housing Department have for a long time
purely wanted to maintain the consultation on the individual,
one-to-one basis, in terms of sending out letters to tenants,
and receiving replies. Of course this is all right where there is
an individual tenant problem, but when it comes to – sort of
– policies, say, on estates, and policies on overall council
housing in the borough, then it provides no route for discussion
on that.'

(Councillor, Sainsborough)

Our evidence, however, suggests that there was no such simple division between parties or between different types of actors. In Braehead, for example, a few Labour councillors expressed sympathetic attitudes towards tenants' groups and tenant participation, but the prevailing tradition in the Labour Group was not to involve tenants directly. The opposition parties supported more contact between the council and tenants. The Leader of the small Liberal Democrat group argued:

> 'I believe that we should lean more towards ... having areas where there are groups of tenants who are then consulted when it comes to things like repairs and refurbishment ... I would like to see if tenants wish to have tenants' co-operatives, that has never ever been discussed.'
>
> (Councillor, Braehead)

And, from a member of the Scottish National Party:

> 'I think the one obvious step that could be made was to have tenants involved in the decision making process, come along to the committees – we do it in (the) Equal Opportunities (Committee).'
>
> (Councillor, Braehead)

But these councillors were unlikely to have much influence over the authority's overall conduct of its relationship with tenants. More influential might have been the (majority) Labour Group Leader who had been responsible for co-opting special needs and women's groups onto the Equal Opportunities Committee, but who had not devoted much attention to housing issues, apart from in his own ward, since being elected to the leadership. The Convener of the Housing Committee, in contrast, had traditional views about his role:

> 'I think that many of the tenants who wish to set up these (tenants') groups don't realise the work that is involved ... I must be quite honest, setting up some of these groups, from time to time, in some instances, you're making a rod to break your own back.'
>
> (Labour Chair, Braehead)

This leading councillor had never shown much enthusiasm for tenant participation, and had never encouraged officers to develop it, other than in making a response to complaints or requests for information received from tenants. The situation in the citizenship councils was different. Senior councillors had played an important role in pushing developments in tenant participation, including the fostering of tenants' associations by means of advice, grants and practical assistance.

Councillors' attitudes to how the relationship with tenants should be conducted could be influenced by the prevailing attitudes and experiences of their authority. Some councillors in the traditional authorities had a more limited view of the possible range of ways the relationship could be conducted, seeing the provision of information to tenants, for example, as the most important, or the only, aspect of the relationship. When asked about the council's policy on tenant participation, one councillor, a member of the minority Liberal Democrat Group on Talybont, said:

'I think it's quite good now. It wasn't very good a few years ago, but the present Housing Manager seems to have . . . full consultation, because we have difficulty with what I call decanting, where we've been doing major repairs, we've had difficulty . . . but I think what they do now, the Housing Department, is more consultation in that they will get all the tenants together, either use the hall or one of our buildings, get them together, and explain in detail exactly what's going to happen, and I think the paperwork that goes out now is very enlightening for the tenants, it instructs them, tells them exactly what procedures are going to take place.'

(Liberal Democrat Councillor, Talybont)

This view of 'consultation', in fact information provision, contrasts with the view of the Liberal Democrat Leader in Braehead, also a traditional authority, quoted above, who saw a range of types of contact and involvement, extending as far as tenants' co-operatives. This councillor was generally better informed about the possible forms of tenant involvement despite the limitations of her own authority's experience. This knowledge may have arisen from her educational background (she was a teacher), or from her active participation in her party's affairs at the national level where she may have been influenced by other Liberal Democrat councillors' experiences and views.

In contrast, in the citizenship authorities there was a history of more far reaching forms of communication and involvement which informed the councillors' views. In these authorities the issue was seen more as being about whether or not tenants' representatives should have voting rights on committees, for example, rather than whether there should be face-to-face contact. Some councillors in the citizenship and consumerist authorities would have preferred a different sort of relationship between the authority and tenants, but they did so from a very different level of knowledge of the possible forms of tenant participation. For example, a councillor in Sainsborough said:

'I'd like to see more (tenant) participation altogether by choice, but I don't necessarily believe – in fact I don't believe – that having tenants, or a tenants' representative or two tenants' representatives on the Housing Committee would actually help anything very much because those two individuals – it's a lot of weight to put on their shoulders to say they would be representative of the whole of the tenant community.'

(Councillor, Sainsborough)

To sum up, in relation to tenant involvement in housing management, the more important division between councillors was not whether they saw themselves as 'tribunes' or 'statespeople', but whether they saw tenants as playing any direct role in the authority's decision-making or management processes in addition to the representations councillors might make on behalf of constituents. Some, including many senior councillors in the traditional authorities, saw it as sufficient for individual tenants to be able to approach their councillor and have him or her take up an issue. Others – a minority of those we interviewed – saw a role which could be described as facilitator. For some councillors, especially in the citizenship authorities, this meant value was attached to tenants having face-to-face discussions with councillors or officers, and these discussions were often to be conducted through tenants' associations. In the consumerist authority the role of the councillor was less clearly delineated, but individual tenants' views were seen as an important component in the council's decision-making processes, and it was not seen as sufficient to rely on the complaints brought to councillors' surgeries. Councillors in the traditional authorities demonstrated a range of attitudes, but tended to have less knowledge of ways in which tenants had been directly involved in housing management elsewhere. Some were satisfied that the traditional role of councillors in taking up individual tenants' complaints was sufficient. Where this view of the role of councillor was held by senior councillors it had a potentially crucial effect on the style of the authority's relationship with tenants.

The implications of this for the typologies of councillors discussed above are that the emphasis on 'tribune' and 'statespeople' roles is over-simplistic. Councillors can be seen to have potentially different attitudes to their role which have crucial effects in defining the nature of relationships with citizens and consumers. In relation to council housing, a simple division can be made between those who see their role as speaking for tenants and those who see their role as that but also enabling tenants to speak for themselves. A further division of the latter group might then be made between those who think tenants should be able to negotiate on behalf of themselves or other tenants, directly face to face with housing managers and councillors; and those who think that individual tenants should be able to have their views taken into account as part of the decision-making process. In other words, traditional, consumerist and citizenship orientations were detected in the attitudes of councillors interviewed, but not necessarily consistently within each authority.

WHY SUPPORT TENANT PARTICIPATION?

We have shown that almost all councillors in the six local authorities believed in involving tenants in some way in the management of council housing although there were considerable differences in their views on the nature of that involvement. The question to be addressed here is why they held this belief. We showed in Chapter 3 that housing managers supported the involvement of tenants for a mixture of selfish, ideological and altruistic reasons. They thought it would make their job easier and help defend their jobs by encouraging tenants to stay with council housing but at the same time had a professional desire to provide a good service to tenants and felt that tenants had a right to be involved in the running of the houses they lived in.

In general the attitudes of councillors were very similar to those of housing managers. Most councillors felt that tenant participation was useful in helping to provide a good housing management service to tenants through the communication of information and views.

'I don't believe that housing departments and councillors know everything and that the input by tenants in many cases is very important. It helps to shape decisions. It can also help to be economic in housing. I think it plays a vital part. I have always been one for consultation.'

(Councillor, Sainsborough)

Most councillors also accepted the view that tenants had a right to be involved in housing management but this was coupled with an awareness that dissatisfied tenants could create considerable problems for councillors.

'We are acting on behalf of them. They are paying for the service. They elected us to do the job yes, but if we ignore their wishes it is at our own peril.'

(Chairperson of Housing, Compton)

'I feel that if tenants are consulted they are much happier about the local housing department and the council. If they are not consulted they feel out of it and bitter. That can cause problems.'

(Councillor, Braehead)

Potential problems were often defined in terms of the perceived threat to council housing from the government policies outlined in Chapter 1: 'In the current climate tenant participation is crucial. With Tenant's Choice it is crucial that they get the idea that the authority is trying to help them' (Councillor, Dockport). Tenant participation, whatever the form, is held by most councillors to enable them to deal with potentially threatening problems before they become serious.

'The benefit (of tenant consultation) is to the Council because then they get the feeling of the people. If there is something going wrong, rather than it being underlying, then it would come out and you can do something about it.'

(Councillor, Talybont)

However, it was also felt by some councillors, particularly in traditional authorities, that tenant participation could backfire on them. Some councillors felt that support for tenants' groups could endanger their position by providing a focus for local discontent and an alternative local leadership which could threaten their position and even lead to tenants' representatives standing for election against them. This kind of view was not a general one even in traditional authorities, and the potential disadvantages of participation were usually seen to be outweighed by the potential advantages.

'If you find tenant participation threatening then you are not a good councillor, because if you are afraid of opposition then you can't be doing your job right. But if you antagonise them

94

(tenants' groups) and don't meet with them then they're going to bring someone forward to challenge you eventually, because you haven't listened to what they're saying. You haven't carried the public with you.'

(Chairperson of Housing, Dockport)

COUNCILLORS' VIEWS OF HOUSING MANAGERS

We have stressed that tenant participation is a tripartite relationship and so it is important to examine the attitudes of councillors towards housing managers. However, the relationship between councillors and officials has always proved to be a difficult one to unravel because it seems to depend crucially on the relationship between individuals. The view that councillors decide policy and officers carry it out is contradicted by the Weberian concern that officers dominate the policy process as well as the detail of administration. In practice it is likely that neither group always dominates. As we shall see in Chapter 7 councillors did not always feel in a strong position to influence policy and sometimes forged an alliance with tenants to push through particular decisions.

All the councillors interviewed in the six local authorities expressed respect for the professionalism of council officers and many thought that most housing managers supported tenant participation. On the whole, councillors felt that housing managers were anxious to listen to tenants and provide a service the tenants thought was satisfactory:

'I know that most of the housing managers . . . know all their people quite well, and tenants who act as a spokesman (sic) in the area, and they will go and talk to them . . . They like feedback, it does make their job easier . . . '

(Labour Councillor, Dockport)

Councillors, on the whole, perceived officers as having a positive attitude to their relationship with council tenants, but there were some differences between traditional and other local authorities, and within local authorities.

Some councillors did not know the officers' attitude to tenant involvement or tenants' associations: 'I wouldn't know to be honest with you . . . So – I think – they may well feel threatened by tenants' involvement' (Councillor, Talybont).

There was less evidence on which to base a view about officers' attitudes in the traditional authorities:

> 'I've never asked our officers what they think (about tenants' associations) so I don't know. But I think in general they are quite happy, quite content to meet council tenants, and welcome their ideas.'
>
> (Councillor, Dockport)

A few councillors had more definite views. For example, a leading Labour councillor in Dockport said:

> 'I think it (tenant consultation) was – um – a bit of a thing for the boss (Director of Housing) to swallow. I think he's more of the "I know what's best, anyway" type . . . he's accepted the inevitable, and I don't say accepted reluctantly, I think it was pushed on him, but once having got it, he does the business the way it should be done, as I understand it. The Deputy is a lot more enthusiastic, because he's new anyway, and I think he's an excellent officer. Because he's full of enthusiasm, and knows that's the way forward for everyone.'
>
> (Councillor, Dockport)

However, the Director in Dockport expressed considerable reservations about tenant participation, and as we shall see in Chapter 6, was heavily criticised by some tenant activists for his attitude towards them.

Many councillors felt that some officers were more sympathetic to the idea of consulting tenants than others. Some councillors felt it would benefit officers to spend more time listening to tenants and being accountable to them, especially if they have 'a bad attitude, then it should be corrected. And if tenants say "'They were rude to me". Well, we need to know that' (Councillor, Talybont).

A few councillors felt that officers may be opposed to tenants' associations:

> 'I think they're completely anti. They would be against tenants' associations, I feel . . . I'm going on an experience in the past. I mean the present housing officers haven't been involved as I know with tenants' associations, but the ones in the past, as soon as the tenants' association was mentioned, they were appalled. They – you know – felt it was a challenge, you see, but I don't think the present housing manager would

96

be concerned about an association, because he is one for communication ... '

(Councillor, Talybont)

The possible significance of these attitudes is demonstrated in Braehead, where one councillor said officers will:

'give the tenants' groups the right answers, but, I think, like everyone else, they probably see these groups as threats, or if they become too much involved in the day to day operations of housing, then it may make their jobs more difficult.'

(Councillor, Braehead)

This perception was based on very little evidence, and officers reported a variety of attitudes in their interviews. Yet in Braehead most officers thought that councillors were virtually all opposed to more tenant consultation or involvement, while most councillors either did not know, or thought that virtually all officers were opposed to it. In our interviews, both groups demonstrated a range of attitudes, from enthusiastic to opposed. In practice, neither group had taken much initiative to develop tenant participation, and in the absence of any organised demands to do so from tenants, little was changing, and neither officers nor councillors felt that the other group had much enthusiasm for tenant participation.

In contrast, in the citizenship and consumerist authorities a councillor when asked what was the attitude of housing officers to tenant participation, replied:

'Well, in my own experience, I think it's a bit mixed. I think some officers welcome tenant participation, the tenant's involvement, if you like, in the everyday management of housing ... On the other hand, there are some officers, I think, who are not very keen on too much participation by tenants' associations or by tenants, feeling that they're imposing on their ground. They're officers; they're managing and what have you the housing stock, the housing department, and they know best, for tenants.'

(Councillor, Compton)

The councillor went on to say that he thought officers in most sympathy with Labour Party policies were most sympathetic to tenants and tenant participation. Another councillor, in Strathmuir, said that officers 'all the way through the structure' believe that 'there should be tenant involvement', and in one improvement project in

97

his ward 'we've had building surveyors involved as well and they appear to be perfectly happy for the tenants to be involved in the working parties' (Councillor, Strathmuir).

A councillor in the consumerist authority, Sainsborough, thought that officers were in favour of tenant consultation, and that 'the District Manager would prefer to have a closer liaison with tenants'. Councillors in this authority showed perhaps the most consistent views about officers.

What matters here is not so much whether councillors were good judges of the attitudes of officers, but whether their views of officers' attitudes influenced their own behaviour. It seems that what councillors thought about officers' attitudes tended to reinforce the prevailing ethos of the authority. There was most overlap between what officers said about their attitudes and what councillors thought officers' attitudes were in the citizenship and consumerist authorities. This may be because senior officers and councillors in these authorities appeared to see more of each other in the company of tenants, possibly providing a fuller opportunity for both groups to gain a better mutual understanding, as well as an understanding of tenants' motivations and attitudes.

CONCLUSIONS

This chapter has reviewed the attitudes of councillors to tenants and tenants' associations, and the perceptions held by councillors of housing managers. The role of councillor is seen as complex and demanding, particularly by councillors. The dichotomy between ward and policy work exists more in theory than in practice, since for most councillors the challenge is to combine both aspects of the job. However, many councillors perceive that one or other role is more time-consuming or demanding. But councillors perceive other conflicts, dilemmas and difficulties in the conduct of their relationship with council tenants. Virtually all see a need to take account of tenants' views, but there are three dimensions or aspects of their approach and attitude to the relationship with tenants on which they may differ with their colleagues, or councillors elsewhere.

The adequacy of the classic 'representative' role

Some councillors thought their traditional representative role was sufficient to allow the voice of tenants to be heard, through themselves or, if more direct participation by tenants was necessary, this was due to factors outside the control of the authority, such as central government policy. However, others believed that this conception of the councillor's role was inadequate, and some actively sought to promote or assist alternative forms of access to the council for tenants.

Most councillors saw it as a legitimate activity for tenants and tenants' groups to attempt to get the councillor to support and argue their case in the council. But a few saw it as important to assist tenants to argue their own case, which is a very different matter. These councillors talked of their role in a way which sounded like the perception community development workers hold of their role – to assist people to take more control of their own destiny, to have a say or present a case as effectively as possible, and to learn how to do it better for the future. This conception of the councillors' role is hinted at in Gyford's use of the word 'facilitator', but the role has not been elaborated clearly.

The potential role of tenants' associations

Some councillors were more supportive of tenants' groups than others. A few were concerned about the representativeness of tenants' associations, or other tenants' representatives. Others were content to see this as a possible concern which needed to be taken into account. At the two extremes of the spectrum of opinion were two (Labour) councillors. The Housing Convener in Braehead felt that supporters of tenant participation did not appreciate how time-consuming tenants' association activity could be, and by implication, felt that the councillor could do whatever was necessary for the tenants (and if he could not it was because the government would not provide the resources). In contrast, a Labour councillor in opposition in Sainsborough felt that tenants' associations were crucial to debates about housing policy at the neighbourhood or authority level, and likened them to trade unions.

The appropriate methods and forms of communication between tenants and landlord

Some councillors in the traditional authorities had very little knowl-edge of the possible forms of communication and involvement between council and tenants. Others, and especially those in the citizenship and consumerist authorities, were better informed through wider contact or personal knowledge. They supported a spectrum of forms of involvement from providing tenants with information, to tenants' co-operatives and tenants sitting on council committees and sub-committees. Their own experience and knowledge seemed important in determining the forms of tenant involvement they supported. Familiarity sometimes bred contempt, however – for example, for the idea of tenants sitting on housing committees. In other cases, it led to a desire to extend and develop new forms of representation for tenants.

In this chapter we have sought to highlight aspects of the councillor's role which have received very sparse attention in the academic literature. In particular, we have examined a perception by some councillors of a dimension to their work which is not conveyed in the classic distinction between local ward and policy dimensions of the role. We found evidence that some councillors perceive a role in assisting access to the council by tenants. Many perceived some difficulties and dilemmas for them in this, but nevertheless felt it was right to accept, if not encourage, the development of tenant participation in housing management. We saw in Chapter 3 that some officers felt that councillors either resisted tenant participation because of the threat it represented to their traditional role, or supported it reluctantly because not to do so would have implications for their relationship with the electorate at election times. Few officers detected in councillors the sort of principled and complex commitment to citizenship or consumerism which many councillors espoused.

5

TENANTS

The aim of this chapter is to look at council housing from the view-point of the tenant. We examine what it means to be a council tenant and how tenants perceive their landlord and the service they provide. The emphasis is on the views of 'ordinary' individual tenants whether or not they are members of, or active in, a tenants' association, and in Chapter 6 we consider the collective activities of tenants through tenants' associations. The empirical material in the chapter largely comes from a questionnaire survey of nearly 1,000 tenants in our six case study areas and from a small group discussion with about a dozen tenants held in each area.

Four issues are addressed in this chapter. The first is the attitudes displayed by tenants towards their landlord, and their feelings about their status as council tenants. The second theme is the extent to which tenants' views can be considered heterogeneous. Are there systematic variations in the attitudes of tenants between different local authorities and between different estates within the same local authority, and, if so, do the variations mean it is misleading to talk about council tenants as a group having common perceptions and attitudes? The third issue is the relationship between individualist and collectivist orientations in the perceptions of tenants, and in their relations with their landlord. This is clearly related to the fourth issue which is the impact of the three types of local authority outlined earlier on the views of tenants. In particular does the adoption of any of these models influence the perceptions and attitudes of tenants and overcome the alienation referred to earlier?

The discussion of tenants' views in this chapter needs to be placed in the context of the changing role of council housing outlined in Chapter 1. Since the 1960s at least there has been a continuing change in the socio-economic profile of council tenants, with lower income groups becoming increasingly concentrated in the sector. In 1953, 16 per cent of households in England and Wales in the lowest quartile of the income distribution were local authority tenants. This increased to 29 per cent in 1965 and 43 per cent in 1976. Similarly, the proportion of economically inactive heads of household has grown among council tenants (Merrett, 1979). 'In 1976 council and housing association tenants' average incomes were just over half those of homebuyers but by 1992 they represented little more than one-third' (Wilcox, 1994, p.46).

Since 1961, local authority housing has increasingly become the tenure of those unable by reason of occupation, income or other circumstances to enter owner-occupation (Clapham and English, 1987, p.39). Successive governments have treated owner-occupation as the 'natural tenure', and there has been a strong ideological push to promote owner-occupation particularly during the 1980s. Financial incentives have been weighted against council housing, and the introduction of the Right to Buy in 1980 provided a new route for many better-off council tenants to move into owner-occupation. The cutbacks in local authority building and improvement programmes and the substantial real increases in rents have made council housing an increasingly unattractive and expensive tenure for those in a position to choose. As a consequence, it is becoming a residual tenure for the poor and the elderly.

In our survey of 1,000 council tenants over half (54 per cent) were in receipt of income support. Only 14 per cent of respondents were in full-time work and eight per cent were in part-time work. The other 78 per cent were not employed for one reason or another with 30 per cent being retired and 27 per cent housewives. Although it is difficult to ascribe a socio-economic group to a population with so little connection to the labour market, 46 per cent of tenants were placed in the lowest socio-economic group (group E). Seventy per cent of those classified as the chief wage earner in households were or had been manual workers and five per cent had never been employed. A similar picture is painted by a survey of 2,464 tenants in 1991. Of these only 24 per cent were in full-time or part-time employment (Bines et al., 1993, p.48). Clearly council tenants are in general a relatively disadvantaged group.

The chapter begins with a review of what it means to be a council tenant in general terms, linking previous surveys with the views of tenants in the six local authorities in this study. It then continues by examining what tenants think about the quality of the service they receive from their landlord, and by looking at the contact that tenants have with their landlord and how they feel about these contacts. A key theme here is the reconciliation of individualist and collectivist approaches to tenant–landlord relations. Finally, tenants' attitudes to tenants' associations and tenant participation are reviewed.

WHAT DOES IT MEAN TO BE A COUNCIL TENANT?

Accounts of what it is like to be a council tenant have tended to paint a picture of a stigmatised group of people. According to Tucker, 'council estates and their people are looked down on as a matter of course' (Tucker, 1966, p.22). Many council tenants are reported as believing that the ill-feeling directed towards council estates by owner-occupiers and private tenants was due to resentment of the perceived subsidy of council housing (Tucker, 1966, p.152). Ward summarised what he called the myth of the council tenant in the 1970s:

> According to a still widely held view, he pays a derisory rent, heavily subsidised by the rest of the rate-payers and tax payers, while his Jaguar is parked outside the door. He has absolute security of tenure: nothing will get him out, and his children will inherit his tenancy. Maybe he no longer keeps coal in the bath: we are more likely to hear that he has colour telly in the bedroom.
>
> (Ward, 1974, p.13)

Not only were council tenants said to experience hostility and feel stigmatised by other tenures, but many also were said to have felt patronised and ill-served by their local authority landlord. As a council tenant, wrote Ward:

> you are the victim of a thousand small humiliations unknown to the owner-occupier, and foreign even to the private tenant, who is never expected to feel that the landlord is doing him a favour.
>
> (Ward, 1974, p.12)

Possibly the social and institutional relationships surrounding council housing are as old as the tenure itself. Some of the earliest council tenants, members of the urban poor, were expected to be passive recipients of council housing. As evidence of subservient status, tenancy agreements in the past stressed the tenants' obligations and rarely mentioned their rights or the landlord's responsibilities. Some tenancy agreements still do this, despite campaigns during the 1980s to attempt to redress the balance. But has this view of the low status of tenants any validity in the 1990s after the granting of new legal rights to tenants in the 1980s, and the increasing emphasis on participation, consumerism and improving performance in the public sector?

The study of *The Nature and Effectiveness of Housing Management in England* (Maclennan *et al.*, 1989) found that there has been little improvement in the relationship between council tenants and their landlords as described by these earlier writers. Tenants felt that council staff did not care about them, were not interested in them, and had no sense of their existence as individuals (Maclennan *et al.*, 1989). Tenants' perceptions of powerlessness and of stigma emerged in their views on the way they were treated by staff:

'The attitude's wrong because it's Victorian. You're peasants and they used to believe round here, because it used to be all Manor places, that a peasant pays his rent and that's it.'

'But then you shouldn't have to get tough. They're public servants, not Gods – that's what they think they are.'

'as I said before, when he first came — was a proper little dictator – I'm the king of the castle sort, you know. You're only a tenant, be quiet.'

(Tenants quoted in Social and Community Planning Research, 1988)

Although tenants did exempt some individual members of staff from this criticism and some landlords fared better than others (Maclennan *et al.*, 1989), all types of landlord were perceived in a similar light. In other words it was not just council housing which was being criticised but landlordism in general. The report argued that a remaking of landlord–tenant relations was urgently required.

Saunders (1990) argues that the dissatisfaction of council tenants with their situation has increased as their reference group has changed:

Tenants now compare their situation with other people who have bought their homes rather than those who rent old houses from private landlords . . . One reason why so many council tenants are unhappy with their housing is therefore that they are no longer grateful simply to escape from poor-quality private rented housing, but are instead increasingly resentful at being trapped within the paternalistic embrace of the local authority.

(Saunders, 1990, p.95)

Saunders claims that the powerlessness of council tenants is at the heart of their dissatisfaction.

the overwhelming impression remains that the basic problem is a crippling and frustrating sense of powerlessness felt by many tenants. Many of those to whom we spoke felt unable to do anything when the clerks in the office ignored them or when the workers who came to their homes did a bodged job, and it was this inability to control factors affecting their own domestic environment which drove so many of them to anger, fatalism or despair.

(Saunders, 1990, p. 94)

Paradoxically the results of household surveys with tenants show more positive attitudes towards staff and the landlord in general than the more qualitative studies seem to show. For example, 75 per cent of council tenants in a 1991 survey said they were very or fairly satisfied with the overall housing service, and only 17 per cent were dissatisfied (Bines et al., 1993, p.65). This is a slightly higher level of satisfaction than that (67 per cent) reported by Maclennan et al. (1989, p.95). Qualitative research techniques tend to elicit negative feelings which are not always clearly apparent from structured questionnaires using simple satisfaction scores although, as we shall see later, many tenants questioned in our survey expressed considerable dissatisfaction with their landlord.

Tenants in our research tended to reinforce the general view of a stigmatised sector:

'My view is why should the working class still be classed as the skivvies. Why should we have to live like this, the middle class don't . . . I can honestly say I don't see why I should have to go to the council when every time I want something doing in my house, say to them, right if you don't get out and do it, I'm going to see my solicitor, because that's the only way I get them to move. They look at me as though I am the dirt

105

of the world and why should we be still classed like that. That
was years ago the workhouses. We're not workhouse people,
we are the backbone of the country as far as I'm concerned
and we should be given our rights, as people, as human beings,
not be cheated and trod upon.'

(Tenant, Dockport)

This general perception of the stigmatised position of tenants was
reflected in some tenants' views of councillors. As we showed in
Chapter 4 only a few councillors had direct experience of living in
council housing, and some tenants questioned their legitimacy and
their capacity to reflect the views of tenants.

'A lot of councillors, they're not greatly concerned about
tenants, they've never had to live on a council estate, they
don't know the problems. A lot of them've got the attitude,
because you live in a deprived area in a council flat, you're
nothing or haven't got the intelligence to be anything and that's
wrong.'

(Tenant, Compton)

Tenants also had a generally negative view of the way they were
treated by staff:

'Some of the staff can be quite abusive, really nasty.'

'Because you're on social security they think you're stupid.'

'They're dead smug. Like they've got their houses and everything. You're just in a council house. They're all smug. It doesn't
affect them anyway.'

(Tenants, Braehead)

There were exceptions to this general view, highlighting the differences among tenants:

'The council, I always find them very friendly, and when you
go to the council offices they're very helpful in that department
and I mean they try and help people with rents and things
like that you know, and I always find them very helpful people
the council.'

(Tenant, Talybont)

The generally negative views that tenants expressed about the way
that they were treated are at variance with the views expressed by
housing managers and councillors, who said that tenants had a right
to be involved in the running of their houses and should receive a

courteous and good service. This support was given in part because it was felt to be in the interest of councillors and housing managers to involve tenants in order to retain them as council tenants. The actual behaviour of councillors and housing managers towards tenants may have moved more slowly away from past practice than their expressed attitudes may show.

It is possible that tenants are reacting to past experiences rather than what happens at present, as we shall see later that contact with the landlord is usually infrequent. Therefore, tenants who are not active in tenants' associations may not be aware of changed attitudes and behaviour. Further, tenants may be reacting not to the attitudes of their landlord, but to the perceived stigma of their situation as council tenants, or to the unattractive character of their house type or neighbourhood.

It is important not to assume that tenants perceive themselves or are perceived as a homogeneous group. There are clear differences in status between estates and between groups of tenants within estates which are reflected in tenants' attitudes. These differences can have complex origins, but some were created in the original design and allocation of council housing as in the 1920s and 1930s, for example. Damer (1974) describes the experience of tenants of 'Wine Alley', a 1930s slum clearance estate in Glasgow. The tenants had their belongings fumigated on moving in and received the unwelcome attentions of the notorious 'Green Ladies' – the Glasgow health visitors. The new tenants were also treated with hostility by neighbouring residents who regarded them as outsiders who had received favourable treatment in being allocated relatively desirable houses. To this day many of the 1930s slum clearance schemes in Glasgow are stigmatised and unpopular.

In the 1960s Tucker (1966) described how 'general needs tenants' looked down on 'slum clearance tenants' partly because of the relatively generous subsidies they received. Twenty years later we found evidence in our research of antipathy between 'established' tenants and homeless families housed under the homelessness legislation. Opprobrium was particularly directed at one-parent families, who have also been victims of government policy initiatives in the 1990s (Department of the Environment, 1994).

Divisions between council tenants partly reflect the hierarchy of status among council tenants. Parker (1985) quotes one tenant in 'Providence' who saw themselves as facing a status ladder which had to be climbed:

'You see the way I look at it is this: once you've got a council place you're set up for life. This is like the first rung of the ladder if you like. You can only go upwards, it's only going to be a matter of time. From here there's only better accommodation they can offer us. As long as we keep up with the rent, in the end that's what they'll have to do.'

(Parker, 1985, p.319)

The status hierarchy of council areas is partly the result of grading schemes which matched good tenants to good housing and bad tenants to bad housing (Gallagher, 1982). Even modern sophisticated points systems, which are largely based on need and minimise the discretion of staff, lead to higher income tenants, who are able to wait for better offers, receiving better accommodation than lower income tenants who are more likely to have to act in desperation by accepting the first offer made to them (Clapham and Kintrea, 1986). Thus, although council tenants may have some similar experiences there are likely to be differences which may influence their attitudes and perceptions of what being a council tenant means.

THE ADVANTAGES AND DISADVANTAGES OF BEING A COUNCIL TENANT

Despite the negative impression of the status of council tenants given by authors, it is clear from the reluctance of many tenants to support the transfer of ownership of their home to an alternative landlord such as a housing association under the Tenant's Choice and Voluntary Transfer legislation that many would prefer to be council tenants rather than any other kind. However, it is evident that most council tenants do not compare their situation with that of other tenants, but with owner-occupiers. Several studies have shown that the majority of households would prefer to be owner-occupiers. For example, Madge and Brown (1981) in a study of young married couples in three areas found that 90 per cent of the sample expressed a preference for home ownership and only five per cent for council renting even though 23 per cent were council tenants.

The preference of many tenants for home ownership is also shown by Saunders' (1990) study in three English towns, although his was only a small sample of 130 tenants. Of these 78 per cent expressed a preference for home ownership and only 19 per cent for council renting. The perceived advantages of home ownership were partly financial, relating to a perception of ownership of an appreciating asset, and partly related to the independence and autonomy which home ownership was perceived as giving. When owners were asked for the disadvantages of owner-occupation, 43 per cent mentioned being responsible for one's own repairs. When tenants were asked for the perceived advantages of council housing, 29 per cent mentioned getting repairs done, the most commonly cited advantage. Twenty per cent of tenants and 29 per cent of owners perceived no disadvantages in their tenure, but 24 per cent of tenants referred to a lack of personal control and 21 per cent mentioned the lack of an asset in return for rent, the two factors where owner-occupation was seen to be preferable. Seventeen per cent complained of repairs not getting done. Therefore, getting repairs done was perceived as the main advantage of council housing and the main disadvantage of owner-occupation. Nevertheless, as we shall see later, many tenants were very critical of the actual repairs service received from their landlord.

The findings of our survey of council tenants reinforce those of Saunders. More than one-third of the tenants cited the freedom from worry about repairs and repair bills as an advantage of being a council tenant. Just over 10 per cent mentioned security of tenure as an advantage, and the proportion doubled among those over 65. Nearly a third (31 per cent) of tenants stated that there were no advantages. The most widely mentioned disadvantage was delay in carrying out repairs and the poor quality of maintenance (17 per cent).

Therefore, repairs were the most commonly mentioned advantage and disadvantage, thus highlighting the importance of this issue for council tenants. How repairs are perceived depends on the actual experience of tenants, which can vary widely between estates. The differences are exemplified by the results achieved by two estates, one in Compton and one in Sainsborough (see Table 5.1). It is worth stressing that the question was an open ended one and replies were later classified. Respondents were allowed to state any number of advantages or disadvantages.

It is clear that in the two estates the pattern of perceived advantages and disadvantages was very different. For example, in Compton estate, repairs were mentioned by nine per cent of people as an advantage and by 26 per cent as a disadvantage, whereas in the Sainsborough estate 48 per cent of people mentioned repairs as an advantage and no one mentioned it as a disadvantage. The differences between the responses would indicate that tenants in the two estates

Table 5.1 Advantages and disadvantages of council housing in parts of Sainsborough and Compton

	% naming advantages and disadvantages	
	Compton estate N = 69	Sainsborough estate N = 40
Advantages:		
Free from worry about repairs	9	48
Availability	6	5
Value for money	7	0
Cheapness	25	33
Security of tenure	12	15
Convenient location	0	3
Quality of accommodation	3	8
Better than private landlord	13	5
Other	19	20
No advantages	30	10
Don't know	7	5
Disadvantages:		
High rents	1	5
Poor quality of accommodation	9	0
Lack of freedom to make changes	3	0
Allocations and transfer system	9	10
Lack of ownership	4	8
Neighbours/noise	8	8
Delays and poor quality of repairs and maintenance	26	0
Other	19	13
No disadvantages	29	58
Don't know	7	0

Note: Percentages do not round to 100 as people could mention more than one advantage or disadvantage.

hold very different views of what council housing means. Further, these views seem to be based, to a large extent, on their own personal experience of the tenure rather than on any general view.

In our survey, overall, the most important issues which tenants thought that they faced as council tenants were repairs and allocations. But differences in priorities emerged both between and within the case studies. For example, nearly half of the tenants (47 per cent) on one estate in Strathmuir said that allocations/transfers/rehousing were the most important issues facing them compared with less than one-fifth (16 per cent) on another estate.

In the group discussions, tenants frequently mentioned poor maintenance and repair services:

> 'I take a very cynical view of this. I think that the council just keeps you waiting, hanging on in the fond hope that you'll do it (the repair) yourself.'
>
> (Tenant, Braehead)

Some tenants felt frustrated by the knowledge that delays in repairs and maintenance laid the foundation for more serious and expensive problems in the future. In this sense there was a perception of a common interest with the landlord in the good upkeep of the landlord's property:

> 'They (the council) should appreciate our views can help them as well. It can help their spending if they listen to the people that have to live within their districts. We can help them monitor their spending a lot better.'
>
> (Tenant, Talybont)

> '(If council officers attended tenants' association meetings) they would then at least know what is going on in the area and what problems is coming to light and if people know that an employee of the council was coming to that meeting, they would then put themselves out and say Oh yes, we'll attend that. I've got so and so I wanna bring up.'
>
> (Tenant, Sainsborough)

Repairs, however, were a recurring feature both in the survey and in group discussions with tenants. It seems likely that a good repairs service may do more than anything else to improve tenants' satisfaction with their landlord and the service provided. The importance of the repairs service as an element of tenant satisfaction is underlined in the findings of a 1991 survey which found, for example, that 74 per cent of tenants who were very satisfied with the quality of the repair work were also very satisfied with the overall quality of the housing service. In contrast, 76 per cent of those who were dissatisfied with the quality of the repair work were also very dissatisfied with the overall quality of the service they were receiving (Bines *et al.*, 1993, p.129).

TENANT SATISFACTION

In our six local authorities, more than two-thirds of all tenants were satisfied with their house or flat as a place to live and with the neighbourhood in which they lived. However, in three authorities, Compton, Strathmuir and Dockport, no more than two-thirds were satisfied, while in the other three, Sainsborough, Braehead and Talybont, more than three-quarters were satisfied with their homes and their neighbourhood. The consumerist authority and two of the three traditional authorities were those where satisfaction was highest. Two of the authorities where satisfaction was lowest were the citizenship authorities of Compton and Strathmuir which had extensive arrangements for tenant participation. Overall 51 per cent of tenants said that they were satisfied with the quality of service provided by their landlord, although only 14 per cent said they were very satisfied. However, the levels of satisfaction varied considerably between landlords. In one citizenship authority, Compton, little more than a quarter (27 per cent) of tenants were satisfied, while in the other authorities about a half were satisfied.

It is difficult to know how to interpret these results. The problems of deriving any meaning from satisfaction scores have been highlighted by Satsangi and Kearns (1992) and Birks and Southan (1992). Expressions of satisfaction or dissatisfaction by tenants are related to individual experiences, attitudes and expectations. These may be influenced by factors such as house type or nature of neighbourhood over which the landlord has no or little control. Satisfaction with a service relates to whatever perceptions and expectations the tenant, not the landlord or interviewer, considers appropriate. Individuals may vary in their expectations and it is unclear how expectations are formed. Do tenants compare what they receive with their own past experiences, with the experiences of others, or images of the service projected by the media, politicians or the service providers? Tenants may not have any experience of landlords other than their own, or they may have little knowledge of a particular service such as repairs because they have not used it recently. Therefore, as Satsangi and Kearns (1992) point out, judgements of service quality are multidimensional and do not relate simply to the service provided. They go on to identify a large number of tenant characteristics (such as age, income, previous housing history) which have been associated with satisfaction in previous studies.

Therefore, the use of satisfaction surveys to judge performance is questionable, even if the results are compared over time or with other organisations, unless the other factors which influence satisfaction are controlled for. This illustrates a general problem with many indicators of performance, which is that it is difficult to be sure that what is being measured is the quality of the service itself, rather than the influence of a large number of factors outside the control of the service provider. Further, satisfaction is dynamic and thus a service consumed over a long time which may have given great satisfaction in the past, can suddenly contain features that bring high levels of dissatisfaction, or vice versa, as expectations and perceptions change.

Therefore, any evaluation of tenant participation in terms of service provision needs to develop a sophisticated series of measurement indicators; be able to interpret them by reference to changes over time and comparisons between similar organisations; and finally be able to distinguish the impact of tenant participation from the myriad other factors which influence the service. No research study has yet come to terms with all these issues and so there are no definitive answers to the questions about the impact of tenant participation on the service provided or on tenant satisfaction.

However, although almost all councillors and housing managers interviewed in our study believed that tenant participation improved the housing management service and increased tenant satisfaction, tenants did not appear to confirm this view, as the adoption of citizenship style tenant participation structures does not seem by itself to transform tenant perceptions of the quality of service offered or their satisfaction with their house or of the neighbourhood in which they live.

These results cannot be used to say that one model of tenant participation is superior to another because of all the problems outlined earlier. Nevertheless, the data show that there is not necessarily a link between tenant participation and a good housing service and tenant satisfaction. It is possible that the practice of tenant participation in Strathmuir and Compton means that the service provided is better and there is greater tenant satisfaction than there would be without tenant participation. It is impossible to tell from the limited information available. However, it is clear that extensive tenant participation along the citizenship model has not been enough to ensure a good service or high tenant satisfaction. In this case at least, the belief of councillors and housing managers that tenant participation leads to the achievement of service objectives and high tenant satisfaction would seem to be at worst misplaced or at best only a partial answer to what may be a complex problem.

Tenants were asked a battery of questions to find out about their attitude to their council landlord. What emerges (Table 5.2) is a

Table 5.2 Attitudes of tenants to their landlord

Statement N = 973	Talybont % (N=98)	Traditional Braehead % (N=98)	Dockport % (N=194)	Consumerist Sainsborough % (N=175)	Citizenship Strathmuir % (N=199)	Compton % (N=209)
				Proportion of tenants who agree		
The council does not care about its tenants	36	43	46	21	45	58
The council is an efficient landlord	61	59	28	80	57	31
The council gives good value for money for your rent	76	47	49	81	42	44
The council listens to its tenants	46	45	41	63	43	23
The council involves its tenants in decision-making	21	10	28	46	25	20
The council does not keep tenants informed	59	61	55	26	66	51

picture of widespread disenchantment with councils and a feeling that their landlord was not interested in keeping tenants informed or consulting them. These findings tally with those of Maclennan *et al.* (1989) who found that tenants felt alienated and frustrated with their local authority landlords. From the responses, it seems that ordinary tenants feel equally negative about the landlord whether or not the council has a traditional approach to tenant participation or has tried to involve tenants and their representatives collectively at all levels of housing policy.

The authority which stands out as having a high level of satisfaction (Sainsborough) had adopted a consumerist approach, consulting tenants individually and restricting the role of tenants' associations to suit its own needs. Yet even here in the group discussion, tenants said that they felt that the housing department only contacted them when there was a problem, over arrears for example. They did not feel that the council was interested in the views of the tenants.

RELATIONS WITH THE LANDLORD

Most tenants appear to have little face-to-face contact with their housing department. Andrews (1979) describes the contact for most residents with officers and councillors as 'intermittent'. In our survey, nearly half the respondents had not visited their local housing office in the last twelve months, excluding visits to pay rent. Less than one-third had visited more than twice. When they did contact the landlord, even in the 'consumerist' authority, tenants could find it difficult to see the right person:

'At one time you could walk into the town hall and see somebody who could do something for you. Now you go to the civic centre and wander about all day and still not find anybody that could help you out.'

(Tenant, Sainsborough)

Landlord-initiated contacts were even rarer. Two-thirds of tenants had received no visit from somebody from the housing department in the last year. In discussion groups, tenants often expressed cynicism about the landlord's reason for contacting them. Andrews has commented that tenants' encounters with councillors and officers are generally concerned with either the remedy of some grievance or with the imposition of sanctions against some infringement of the rules:

The resident is cast in the role of applicant for benefits of some kind or other or in the role of the guilty party; the officer as dispenser or arbiter; the councillor as advocate or mediator.

(Andrews, 1979, p.225)

It could be expected that the adoption of collective forms of tenant representation through tenant participation structures in the citizen-ship, and to a lesser extent, in the traditional authorities would also alter this pattern of contact. However, in all six authorities most tenants acted initially on an individual basis in their contact with a local authority landlord. In our survey, more than three-quarters of tenants (77 per cent) agreed that it was up to a tenant on their own to make their views known to the council. There was little difference between members and non-members of tenants' associations or between tenants in different types of authority.

Despite this individualistic orientation, tenants in Dockport and Braehead distinguished renting a house from renting a television. Whether this is due to a perceived qualitative difference between a house and other goods, or to an ideological standpoint about the nature of council housing is not clear.

'You can't take a house back. We are paying rent to live in that house.'

(Tenant, Dockport)

'You cannae trade it in for a brand new model at your convenience.'

(Tenant, Braehead)

Although tenants' primary relationship with their landlord was an individual one, they do not necessarily see that as placing them in a similar relationship to that of the consumer in the marketplace. This was demonstrated in tenants' attitudes towards collective action. A very large proportion of tenants (83 per cent) agreed that tenants should get together on an estate to improve their situation because they have a strong common interest. Not only do the majority of tenants perceive a common interest with other tenants at the estate level, but we occasionally encountered tenants and people who had bought their home from the council forming joint residents' associations, particularly in blocks of flats where tenants and leaseholders perceived shared interests in matters such as care-taking and cleaning of common areas (Cairncross, Clapham and Goodlad, 1993).

A smaller majority of tenants (60 per cent) agreed that tenants needed to meet with tenants from other areas to improve their situation because they were all in the same position. The weaker perception of common interest beyond the estate level is understandable, given the variation in experiences and attitudes to the landlord which the survey revealed across estates within the same authority (for a full review see Cairncross, Clapham and Goodlad, 1993). The widely recognised status differences between estates may also contribute to the weaker sense of affiliation between tenants from different areas.

Tenants in the traditional authorities were more likely to agree with the view that it was up to tenants on their own to make their views known to the council and less likely to agree that tenants should meet with those from other areas to improve their situation. It seems that there may be some relationship between experience of some kind of tenant participation and the perception of common interests among tenants. Nevertheless, tenants' attitudes seem to be deeply ambivalent between individualist and collectivist orientations and this is reflected in the action which tenants take to influence their landlord.

When asked what they would do if the council was proposing to do something important which they thought unjust or harmful (see Table 5.3), tenants most commonly mentioned contacting their councillor, signing a petition or contacting their MP. Some actions were considered to be a lot more effective than others. The replies show that a collective orientation is not always converted into a positive attitude towards collective action. People said that they would generally deal with any problem themselves by contacting an influential person such as a councillor or MP. Few tenants said they would have recourse to collective action by, for example, contacting a tenants' association. However, only a quarter of members said they would raise the matter in their association although, as one would expect, association members were more likely to do this. It seems that members use their tenants' association as one of a variety of means of raising issues where an initial contact with the housing officer has been unsuccessful. In other words, they adopt a pluralistic approach to solving problems rather than choosing in principle between individualist or collectivist approaches. Further, few thought they would have recourse to militant action such as a rent strike or protest action, or believed that such action would be effective.

Table 5.3 Actions in response to unjust and harmful actions of the council

	Percentage who said they would take this action	Percentage who said they would be effective
Contact my councillor	42	31
Sign a petition	41	26
Contact my MP	40	46
Contact a council officer	32	16
Contact local media	14	28
Speak to an influential person	13	12
Stop paying rent	10	14
Form a group of like-minded people	9	8
Go on a protest	7	8
Raise the issue in a tenants' association	6	3
None of these things	8	10
Don't know	5	1

Note: Percentages do not add to 100 as respondents were able to give up to three responses. Total cases = 984.

In group discussions, tenants said they had usually contacted their local councillor or MP or even used a solicitor to deal with persistent housing problems.

'They won't listen to the ordinary lay person. It has to be someone with a bit of authority that tells them where they're going wrong. They ignore us.'

(Tenant, Talybont)

Tenants who are not tenants' association members are more likely to use the traditional instruments of representative democracy: councillors and MPs. Tenants' association members use a greater variety of methods both individual and collective.

The faith in traditional democratic mechanisms is shown in replies to questions designed to measure political efficacy and trust (Table 5.4). These measures have been used to analyse the extent and mode of political participation (Almond and Verba, 1965). Tenants in traditional and largely rural Talybont were less likely to agree strongly with the statement that 'those we elect as councillors lose touch with

118

Table 5.4 Trust and efficacy of council tenants

	Talybont %	Traditional Braehead %	Dockport %	Consumerist Sainsborough %	Citizenship Strathmuir %	Compton %
Parties are only interested in people's votes, not their opinions	75	75	75	72	88	81
Sometimes politics and local government seem so complicated that a person like me cannot really understand what is going on	78	66	74	76	89	82
Generally speaking, those we elect as councillors lose touch with people very quickly	60	67	70	66	79	78
People like me have no say in what the council does	72	77	71	58	82	68
I don't think council officials care much about what people like me think	54	73	67	45	80	73
People in the council waste a lot of the money we pay in rates	54	66	63	52	76	66
Most of the time you can trust the council to do what is right	49	38	39	53	40	28

Note: Total cases = 984.

people pretty quickly.' In general, they had a higher sense of political efficacy, being more likely to believe that they understood political discourse and had a say in what local councils did. Also, they were more likely to trust the council to do what is right.

Differences between local authority estates on these measures were marked. For example, the proportion agreeing that councillors lost touch varied between 51 per cent in one estate and 87 per cent in another in a different local authority (41 and 65 interviews), perhaps reflecting differences in the behaviour and accessibility of councillors. Trust in the local council to do what is right varied from 25 per cent agreeing in two estates to 59 per cent in another (83 and 48 interviews, and 44 interviews respectively).

In general tenants' responses reveal a situation where the majority of tenants feel politically ineffective and distrustful of their local authority. There is no sign that those in the citizenship authorities felt more effective in the political sphere than elsewhere, and they appeared to have no more confidence or trust in their local council than those in other kinds of authority.

This suspicion and cynicism was echoed by some tenants in the group discussions:

'The only time you see a councillor is at the elections and they come knocking at everyone's door.'

(Tenant, Talybont)

'They (housing staff) look at me as though I am the dirt of the world.'

(Tenant, Talybont)

Tenants believed that the relationship with the landlord should be a reciprocal one. In many cases, tenants did not feel that this was the reality.

'If the tenant carries out his or her duty, the council should commit themselves and carry their duties out.'

(Tenant, Talybont)

'We have to fight and fight and fight until we're listened to and you get fed up and you get disheartened.'

(Tenant, Talybont)

Tenants in some authorities felt that while the council expected tenants to fulfil their obligations, such as paying rent, it was difficult to get the council to meet its obligations to carry out repairs. If you complained about delays, then you would have to wait even longer. For example, in traditional Talybont tenants felt that their relationship with the council landlord was one-sided:

'If we're paying (the local authority) our rents and rates or whatever, they should surely answer to us, not us answer to them. We're paying them to do a service for us, so surely they should be listening to us and we should be in control of them.'

(Tenant, Talybont)

ATTITUDES TOWARDS TENANT PARTICIPATION

Some tenants were cynical about the reasons why their landlord pursued tenant participation. They thought that the council would consult them and then do whatever it had originally planned to do, regardless of the results of the consultation.

'If they (the council) sent a survey out – what do you want doing? – I'd guarantee we'd all put 10 different things down probably. Those surveys they throw on one side and they do what they want to do, not what you want.'

(Tenant, Dockport)

Tenants involved in tenants' associations could be equally cynical about their landlord and the reality of consultation. Both the quotes below come from a tenants' association member in a citizenship authority.

'Twice this year I have walked from the top to the bottom of every tower block with the Chair of Housing and I've got a list of things that big that were written down which he sent to the housing officer that needed attention. The only thing that's been done is the windows and that's all that's been done.'

(Tenant, Compton)

'We had a meeting of the tenants (about faulty lifts) where people came from the council, so in a sense, we, you could say that they involved us in consultation, but all they could say (was) "Well we accept responsibility" and they gave us a date when things will be done by and, of course, the date passes and the thing's still not working. So you can say what you want

121

and you can be consulted as many times as you want but if they don't do it, it doesn't really get very far.'

(Tenant, Compton)

There was also some scepticism about the local authorities' ability to respond to tenants' views even if they wanted to:

'You know it's a bit of a bogus procedure for them to involve us in consultation and discussion about what we want to see when they know full well they haven't got the money, you know and it's just taking the mickey out of us really.'

(Tenant, Compton)

However, tenants thought that they had a right to be kept informed and consulted by their landlord. In the survey, most tenants (60 per cent) felt that their landlord provided them with too little information. Two of the authorities where tenants were most likely (69 per cent and 72 per cent) to say that the council gave them too little information as tenants were citizenship authorities, while the one authority where the majority of tenants appeared to think that they received enough information from the local authority was the consumerist authority.

In group discussions, tenants expressed a wish for more information and consultation from the landlord. They justified more tenant involvement on the basis of their perceived rights as tenants in the same way that housing managers and councillors did: 'The tenants have got to live in the houses, so they've got to be consulted. It's their houses, you know, that you're talking about' (Tenant, Dockport).

Less than one-fifth of tenants thought that in general their council found out about tenants' views before it took important decisions which affected them. The variations in tenants' responses to this question across the six authorities were considerable. In Strathmuir and Compton, 9 and 16 per cent of tenants respectively thought that the council found out their views; in the consumerist authority, Sainsborough, 37 per cent of tenants felt the council found out their views, and in the three 'traditional' authorities 10, 13 and 18 per cent of tenants said that the council found out their views before taking important decisions which affected them. The advocates of a citizenship approach would argue on ideological grounds that tenants should have a right to have a say in decisions affecting them. The results of our survey indicate that, where such an approach is being put into practice, tenants' perception of this is little different from that of tenants in other types of authority. If anything, it is worse than in the consumerist authority. It seems that there is no particular relationship between a citizenship approach to tenants and tenant participation and a feeling among tenants that their views are considered by the local authority when making decisions affecting them. In Compton, for example, it was felt that tenants who were not involved in the participation machinery were not kept informed by the local authority. In this authority, district committees which were composed of tenants' association representatives and attended by councillors and housing staff were regarded as a key source of information. This underlines the important role of tenant participation as a means of communication between landlord and tenants and also raises the question of how to keep tenants informed who are not active in the arrangements for tenant participation.

When tenants were asked how they would like the council to ask for their views about its policies and plans, the most frequently mentioned method was by individual letter (45 per cent) followed by public meetings, visits from a housing officer, and newsletters (Table 5.5). Other forms of consultation were less widely mentioned. Tenants' association members were much more likely to mention consultation through tenants' representatives as a preferred means of

Table 5.5 Methods for informing tenants about housing policies and plans

Method	% of tenants supporting method
Advert, poster or notice	3
Leaflets	14
Newsletter	24
By writing letters to tenants individually	45
Through surveys like this	13
Directly from councillors	6
Directly from housing staff in housing office	10
Visit from housing officer	26
Through tenants' representative	12
Public meetings	29
Committees with tenants	4
Local papers or radio	3
Don't know	13

Note: Total cases = 984.

consultation than other tenants. However, even this group were more likely to say that they would like the local authority to ask them for their views through individual letters than any other method.

Thus it seems that, whether or not tenants are involved in a tenants' association, their primary perception of their relationship with the landlord is an individual one. However, the responses to these questions indicate that a significant proportion of tenants also value the opportunity for a more collective relationship with their landlord.

More than three-quarters (77 per cent) of respondents in our survey wanted to have more say in the way their housing was run and rated 'the right to have a say in the decisions that affect your home' as very important. The percentage was lowest in the consumerist authority, although even there it was more than half. There is a widespread feeling among council tenants that they would like to have a say in decisions about their homes, and the findings of our survey indicate that for most of them existing arrangements are not satisfactory. What is interesting is that tenants were more aware of efforts to consult them individually than collectively.

Very few tenants (8 per cent) agreed that councillors were the only people who should make decisions about council housing and that tenants' associations should not get involved. Thus the majority of tenants do not appear to be satisfied with the traditional local authority approach. They think that others apart from councillors should have a say in decisions about council housing. These views could accommodate both the consumerist and the citizenship approach.

ATTITUDES TOWARDS TENANTS' ASSOCIATIONS

Tenants in the survey were provided with a list of positive and negative statements about tenants' associations. Most tenants agreed with the positive ones and disagreed with the negatives (see Table 5.6).

The positive view of tenants' associations was applied across a wide range of issues. They were seen as effective vehicles for getting the council to do things on an estate:

> 'They (tenants' associations) do actually try and get somewhere with the council and the councillors to do things that need doing. And I think they're a great help.'
>
> (Tenant, Sainsborough)

Table 5.6 Tenants' attitudes towards tenants' associations

	% of tenants who agree
Tenants' associations welcome new members	69
They are a way to get the council to do things on an estate	69
They help to create a community spirit	67
They help communication between tenants and the council	67
They provide a place to meet people and make friends	66
They can help with individual tenants' problems	66
They give people experience of working in committees	62
You get something out of going to them	48
They can't change anything	29
They're run by a small group of people out for their own ends	26
They stop tenants from taking more direct action	23
They don't represent what tenants think	20
They're too political	19

Note: Total cases = 984.

'The tenants' association can do and is whatever the people come along and want it to do really ... You know I think the tenants' association has done a lot which a lot of tenants actually don't realise is down to the tenants' association. Like lifts, like repairs done, like getting the council along to see the state of the rubbish and you know dealing with perhaps a caretaker that's not really doing the job and things like that.'

(Tenant, Compton)

There was widespread recognition among tenants of how the existence of a tenants' association could enhance their influence over the local authority landlord. This was even mentioned by tenants in the 'consumerist' local authority.

'As an individual ... it gets forgotten. But collectively, as a group, it doesn't get forgotten, not so quickly anyway ... It has a little bit more push too, behind it.'

(Tenant, Sainsborough)

Less than a third of tenants agreed with negative comments about associations. The most common feeling was that they could not change anything perhaps because of the perceived unwillingness of the council to listen or its ability to respond to tenant demands because of the constraints it faced.

Cynicism towards tenants' associations was shown by about a quarter of tenants who agreed that they were run by a small group of people out for their own ends who did not represent what tenants thought. Some associations were perceived as being too political. Although all of the associations in the six areas made every effort to remain politically neutral in party terms, some of the activists were involved in political parties and had political affiliations which were well known to tenants. The criticisms of political activity may also reflect a view by some tenants that some associations concentrated too much on general policy issues rather than being concerned with day-to-day issues on the estates: 'I think local tenants' associations should concentrate on local issues ... it's not worth joining unless it is going to concentrate on what's local' (Tenant, Sainsborough).

In general tenants showed a positive attitude towards tenants' associations and recognised their usefulness. Where there were no tenants' associations, tenants were eager for information about what a tenants' association would do. In the discussion groups, it became clear that the lack of tenants' associations in an authority did not mean that tenants were satisfied with their housing and the housing service, nor did it mean that they were not interested in the idea of a tenants' association.

CONCLUSIONS

Four questions were identified at the beginning of the chapter, the first of which concerns the attitudes of tenants towards their tenure. It is evident from a number of studies that many tenants compare their own position unfavourably with owner-occupation. In particular many are concerned with the powerlessness of their position in relation to their landlord and the poor levels of service, particularly in repairs, which they perceive as being provided. However, these attitudes vary between different landlords and are not confined to council housing, but apply to housing associations as well.

The second issue was the heterogeneity in the attitudes of tenants between different estates. This emerged in particular in relation to tenant attitudes towards repairs. Having your repairs carried out was perceived as the major advantage of council housing and yet failure to get repairs done was mentioned as the major disadvantage of council housing. Whether repairs were viewed as an advantage or disadvantage depended on the service provided in particular estates. Coupled with the status variations between estates, the different experience of the service provided by the landlord makes it potentially misleading to generalise about council housing as a whole. It also makes problematic the organisation of tenants at a level above that of the individual estate (for fuller discussion of this issue see Cairncross, Clapham and Goodlad, 1993).

Tenant mobilisation is also influenced by the coexistence of individualistic and collectivist orientations by tenants. The apparent contradiction between the two outlooks is easily resolved through the realisation that most tenants were not ideologically attached to one or the other approach. Rather, tenants would use whatever approach seemed appropriate and likely to be successful in a given situation. Therefore, although tenants expressed faith in the effectiveness of tenants' associations this did not stop them contacting their councillor or MP if they thought this would achieve their aims. Tenants had an instrumental view towards methods of involvement and neither the consumerist nor the citizenship view was considered to be superior *per se*. Tenants are likely to judge different ways of involving them by their impact on the way they are treated and the service they receive.

Finally, the evidence of our research suggests that the alienation of tenants has not been overcome in either the traditional or the citizenship authorities. The adoption of collective methods of tenant involvement in the citizenship authorities has not created favourable tenant attitudes towards the landlord. There appear to be two reasons for this. The first is that very few tenants are actively involved in the tenant participation structures and so do not identify with them or know about the issues raised or what is done about them. Therefore, most tenants are denied access to the developmental benefits of participation which will be described in the next chapter. However, the instrumental orientation towards tenant participation means that tenants will judge it on the basis of its impact on the service provided. The two citizenship authorities were judged as providing a very poor service and so tenants were sceptical of their landlords' attitudes towards tenants. But it is not clear that tenants' perceptions took account of the difficulty of the task faced by those landlords. In contrast, the consumerist authority was felt to provide a good service and so tenants were less alienated. However, this evidence does not show the superiority of the consumerist model as it is equally unclear that the perceptions of high quality service reflected any better service than was provided by other authorities. The evidence points to the importance of the quality of the service provided in shaping the landlord–tenant relationship for most tenants, rather than to the existence of any structures for participation.

6

TENANTS' ASSOCIATIONS

Tenants' associations in council estates have a long and poorly documented history. Associations are known to have formed in many estates in the interwar years, and there is some evidence that their number grew in the early postwar years as the council housing stock expanded. In the last twenty years, tenants' associations are thought to have been formed on a larger scale than ever before, as the legal status and rights of tenants grew, and in particular as more local authorities supported their formation as a means of encouraging tenant participation in housing management. The growth in the number of associations was reflected in the formation of federations of tenants' associations in many cities in the 1970s and 1980s; in the provision of grants and other resources by sympathetic local authorities; and in the establishment of organisations such as the Tenant Participation Advisory Services (TPAS) and the Priority Estates Project (PEP) to assist tenants to play a more influential role in housing management.

In the 1990s, tenants' associations have been seen increasingly as the focus for new developments in housing management such as estate management boards and management co-operatives, and tenant participation in the process of competitive tendering of housing management services. Training for tenants' association activists interested in playing a greater role in housing management is provided by a number of organisations (Scott et al., 1994).

It might be thought, therefore, that the role of tenants' associations as an intermediary in the relationship between tenants and councils is unproblematic, developing steadily towards greater and greater involvement. A reading of some research reports, and literature produced by promotional agencies such as TPAS and PEP would suggest this is the case. But academic writing on tenants' associations and housing policy suggests that tenants' associations have been unsuccessful in defending tenants' interests in the face of cuts in the council housing programme, rising rents and increasing concentrations of poverty amongst council tenants. Further, it has been suggested that tenants' associations have changed from being a potentially significant social movement into a local focus for welfare activities; and become incorporated into local authority structures for running community facilities, or estate services (Lowe, 1986).

This chapter begins to answer the questions implicit in these contrasting views of the purpose and impact of tenants' associations. Are they to be seen as in the vanguard of popular opposition to injustice in housing and urban policy? Or are they social clubs, providing for some of the social service and recreational needs of their members and, at worst, becoming incorporated into decision-making structures or processes which subvert their capacity to defend their members' interests? Or should they be viewed in some other way? Evidence from case studies and other sources about the role tenants' associations play in their locality and beyond is examined. This is done from the perspective of tenants themselves, drawing on a postal survey of associations, group discussions with tenants and interviews with tenant activists. The chapter, therefore, provides some background necessary to consider the significance of the growth in arrangements for tenant participation, and the issues of power and influence addressed in the final two chapters of the book.

The chapter starts with a review of the literature on tenants' associations, and goes on to discuss the reasons why tenants start new associations or join established ones. The chapter looks at the activities of tenants' associations in greater detail. Then it looks at the internal organisation of associations, from the 'ordinary' and active member's perspectives. Finally, it examines the specific issue of the relationship with councillors and housing managers.

Research on tenants' associations can be divided into two categories. The first contains work mainly carried out between the 1930s and 1950s into the social structure of council estates (see, for example, Durant, 1939; Morris and Mogey, 1954 and Lupton and Mitchell, 1954). These studies were focused on newly-built estates and concentrated on the 'settling-down' process and overcoming the hostile reaction of host communities. The impetus for setting up a tenants' group is supplied by the need to band together, both to put pressure on the council over management and environmental problems and also to provide the support to counteract the hostility of the host communities. In addition, the association provided a focus which could lead to social networks being built and the transformation of the association into a socially oriented group.

The second category of research studies contains those carried out in the 1960s and 1970s on militant tenants' action about the level of rents (see for example, Hampton, 1970; Moorhouse *et al.*, 1972; Sklair, 1975; Lowe, 1986). There was particular activity over the introduction of 'fair rents' for council tenants in the Housing Finance Act, 1972 which many Labour local authorities were uncertain over whether or not to implement. The tenant campaigns, which sometimes went as far as rent strikes, failed to prevent all but three local authorities from implementing the legislation, but it could be argued they helped build support for the rapid repeal of the legislation on the election of a Labour Government in 1974. Lowe (1986) and Hampton (1970) have charted the large increase in the number of tenants' associations and the amount and militancy of tenant activity around the rent issue from the late 1960s onwards in Sheffield. Many new associations were set up and attempts were made to form a Co-ordinating Committee to fight the rent increases. However by the mid-1970s, the Committee had disintegrated, protest died out and a large number of associations ceased to exist. But, as Lowe makes clear, this did not end the tenants' movement in Sheffield.

> Many individual associations continued to be active, although long term survival was due to a range of factors only partly connected with the rent issue. This frequently involved the incorporation of the associations into management responsibility for a tenants' hall. Of the surviving associations, 60 per cent mentioned in interview in the mid-1970s the management of a hall as their most important current function.
>
> (Lowe, 1986, pp.98–9)

Lowe then builds on his own evidence and that of the previous research on new estates to put forward a general view of the origins and lifecycle of tenants' groups. The stimulus for the creation of groups is held to be an external threat, either in the form of rent increases or, on the new estates, service problems and the hostility of existing residents. He quotes Durant's observation that 'antagonism from without breeds association from within' (Durant, 1939, p.21).

Once formed, it is argued that groups tend to follow a similar pattern of development or lifecycle.

Associations may develop from early agitational activity to a settled routine of social events; this transition may be accompanied by internal faction fighting among committee members; the established functions may include a trade union role of advocacy and bargaining on behalf of the membership as well as social events . . .

(Lowe, 1986, p.101)

Lowe's study and those of others were restricted to particular areas at particular times. For example, in the 1950s and 1960s most of the studies were on new estates. In the 1970s interest had focused on rent campaigns and concentrated on experience in Sheffield in particular. It is difficult to justify a general theory of tenants' associations based on such a limited empirical base.

The approach of authors such as Lowe has been based on the work of Castells and others on urban social movements. Adoption of this framework has led to an over-emphasis on the 'agitational activity' such as rent campaigns designed to achieve important political and policy objectives, at the expense of what Castells derides as 'consumer trade unionism', namely a concern with representing the estates in discussions with statutory and voluntary bodies about services and facilities, and in achieving policy change at the local level. This has resulted in a tendency to neglect the social and welfare activities of tenants' associations which has in turn distorted the picture of the nature of tenants' associations. This can be illustrated with evidence from our six areas.

REASONS FOR FORMING TENANTS' ASSOCIATIONS

The view shared by many commentators is that often 'the spur to initiating an association is a grievance against the authorities about the management of the estate or lack of facilities' (Lowe, 1986). Mayo, in a study of estates in Southwark, found that on 90 per cent of the estates surveyed, the tenants' associations had started because of inadequate facilities or maintenance (quoted in Craddock, 1975). Craddock's own findings indicated a broader range of reasons for formation, from the need to ensure representation on participation committees, to 'working towards educating tenants to manage their own estate'. The Scottish Consumer Council also found that the reasons for the formation of tenants' and residents' associations were

132

varied and often multiple. In contrast to the previous writers, the Scottish Consumer Council (1981) study did not find maintenance problems and other grievances such a common explanation of how a group came to be formed. The promotion of community spirit (18 per cent) and the need to provide a link between landlord and tenant (20 per cent) appeared to be the two most important reasons for formation cited by tenants' and residents' associations.

In our own study, although repairs, maintenance, modernisation and rehabilitation matters were the most common reasons given, almost exactly the same proportion of associations as in the Scottish Consumer Council study mentioned social reasons and the need to establish a channel for contact with the housing department.

Lowe's thesis that most tenants' associations start out as protest associations might lead one to expect a number of the respondents would have mentioned campaigning around issues such as rent rises, the 'poll tax', and proposed housing legislation as reasons for their formation. Although several groups stated that these were among their activities, it is interesting to note that no tenants' association mentioned these issues as having led to their formation. For long-established tenants' associations there may have been some difficulty in recalling the reasons they formed. In Strathmuir, for example, some officers perceived the high rent rises of the early 1980s as being the trigger for many new tenants' associations to form, yet associations' own responses did not reflect this.

Some of the reasons mentioned were unique to one area, while others were only mentioned by tenants' associations in Compton and Strathmuir, where the local authority actively encouraged their formation. It was only associations in these areas that stated they had formed as a result of a local authority officer's initiative. Dampness and anti-social tenants were cited by tenants' associations in only one area as reasons for their formation. Local characteristics, therefore, play some part in determining the issues around which a tenants' association may form.

Social and welfare reasons and the wish to develop a community spirit in an area were mentioned by a considerable number of tenants' associations in the citizenship areas where they received support from the local authority, but by none from other areas. The tentative conclusion from this is that tenants' associations form as 'agitational' pressure groups in all types of area whether or not they receive support from the housing authority, but in areas where they are encouraged (citizenship authorities), they will form for other reasons too.

Associations formed in different time periods reported variations in their reasons for formation. For example, five of the eleven associations formed before 1970 (45 per cent) said they were formed for social or welfare reasons compared to 24 per cent of those formed in the 1970s and 23 per cent of those formed in the 1980s. However, this does not necessarily mean that more associations were formed for social and community reasons before 1970, because many may have ceased to exist. It may be that associations formed for these reasons are more likely to remain in existence.

Another way of looking at the motivations of tenants in relation to associations is to examine why they join an existing association. Tenants join tenants' associations primarily for instrumental reasons: 'to get something done on the estate because I felt that someone should speak up for the tenants'. Common reasons were to campaign for estate modernisation, or social facilities for tenants, or to get to know other people on the estate.

'The services have got worse and worse and, you know, it just seems to me . . . unless you actually organise to fight against that, they'll just, you know, get away with it . . . you can actually achieve things and the council listens if they feel that there's a lot of people against them 'cos who's going to vote for them next time?'

(Tenant, Compton)

The instrumental reasons are highlighted in Table 6.1. Most of the responses were general rather than specific, leaving uncertain the nature of the issues with which tenants would consider an association able to help. Of tenants who were members of their local tenants' association, the most common reasons for joining were to get things done; to get tenants' problems sorted out; to protect tenants' interests; to have more influence as a group; and to get information. Only a few said they had joined for social reasons or mentioned the development of community spirit as a reason for joining (see Table 6.1). The social and community role of tenants' associations is one which may exist independently of the approach to, and arrangements for, participation: 'We like to organise things, the discos and the toddlers club.'

The reasons tenants gave for becoming active were similar to those for joining an association in the first place. They wanted to improve conditions on their estates and to build a community spirit. They believed that a tenants' association gave tenants more power to influence these things.

134

Table 6.1 Reasons for joining tenants' association

To get things done	28%
To get tenants' problems sorted out	17%
To protect tenants' interests	16%
More effective/influence as a group	13%
For information	13%
Political reasons	13%
Was asked to join	9%
A good idea	8%
To get community together	7%
To communicate with the council	7%
Social reasons	6%
Modernisation/rehabilitation	6%

Note: N = 199.

'You're stronger in a group than you are as an individual, so if problems do occur and you run a group, it's helpful, whereas one person, you know, on their own doesn't really get very far. So I think having the backing of a tenants' association is good, because I think the officers take more notice.'

(Tenant Activist, Sainsborough)

The general motives of altruism and wishing to see conditions improved were reflected in tenant activists' views of the role of tenants' associations:

'Mainly to keep everybody happy I think. If you have a happy little community then I think you can feel that things are going well. Obviously to liaise with the council, that's a very important part of it and to try to see their point of view.'

(Tenant Activist, Sainsborough)

'Mainly representing tenants' views, assisting, advising even, tenants in matters, you know, pertaining to their tenancy, welfare benefits, anything.'

(Tenant Activist, Sainsborough)

Tenant activists identify a range of roles for tenants' associations. The social role, a representative role and a wider community role were all mentioned:

'(Role of TA is) to represent the interests of tenants to the

135

local authority . . . (and) they should broaden their horizons and go into the social aspects of the community.'

(Tenant Activist, Strathmuir)

Tenants involved in associations in the citizenship authorities were more likely to mention a social, welfare and community development role for associations:

'We're a group of people who don't know each other, and we've got to learn to get to know each other, to live with each other, like they did in the street I used to live in.'

(Tenant Activist, Compton)

It is already clear that Lowe's account of the typical lifecycle of a tenants' association – 'early agitational activity' followed by a 'settled routine' of social events and 'trade union' activity – is at least too simplistic. Before drawing fuller conclusions, it is useful to examine the activities of tenants' associations in greater detail.

ACTIVITIES

The main activities of tenants' associations were social and welfare activities, but many saw a role as a pressure group. While many groups were formed to campaign around a limited number of issues relating specifically to members' living conditions, this role is replaced to a certain extent with the development of social and welfare activities and the establishment of contact with the housing department once they are set up. Informing tenants also appeared to be an activity of some importance, although it was mentioned by only one tenants' association as a reason for formation.

There were some activities which applied only to the authorities with extensive participation structures. Only one tenants' association outside of the two citizenship authorities mentioned activities relating to the development of community spirit and social and welfare activities. Tenants' associations in one citizenship authority, Strathmuir, were alone in listing work relating to co-operatives, dampness and allocations as among their main activities.

Many tenants' associations have very limited resources. Although one association in the six authorities had a budget of over £10,000 (due to a very successful bar), and a few others had received large grants for specific projects, nearly one-third had an income of less than £50 a year. Only 68 per cent of associations had access to a meeting room, 61 per cent to a typewriter and 45 per cent to a photocopier. Only 21 per cent had access to office premises where organisational activities could take place. Associations in the citizenship authorities tended to have higher incomes and access to more resources, because of support from the local authority.

Our research indicates that tenants' associations are engaged in a broad range of activity, playing both a pressure group and social and welfare role, but often doing so with very limited resources. As well as the two types of activity identified by previous writers on the subject, the survey also reveals the importance of communication for many associations, both with their council landlord and their members. Contact with other organisations, such as the local community council or parish council and, to a lesser extent, with other tenants' associations, was also mentioned by several associations. Political activity around national issues was mentioned relatively infrequently by tenants' associations.

LOCAL AUTHORITY INFLUENCE

The policies and actions of the local authority play a key role in the formation and activities of tenants' associations. An important way in which a local authority can assist the development of tenants' associations is through the provision of resources. In the citizenship authorities, associations received community work support, grants, help with premises and other forms of support. This is still relatively unusual. In 1986–7, more than one-third of local authorities provided no support to tenants' groups at all. None of the Welsh authorities responding to the national survey of local authorities in 1986–7 provided premises or grants to tenants' associations and only one in ten employed specialist staff to work with tenants' groups. Nevertheless, as shown in Chapter 1, the number of authorities in England providing support to tenants' associations had increased considerably so that in 1991 half provided free premises compared to 13 per cent in 1986–7.

A discernible pattern emerged in the nature of tenants' associations in different types of authority. In authorities which were hostile to or unenthusiastic about associations, tenants' groups were more likely to be formed in response to an external threat and to pursue 'agitational activity' than associations in a more supportive environment. None of the associations surveyed in the traditional or consumerist authorities mentioned social and welfare reasons or the development of community spirit as reasons for formation in contrast to a substantial number of those in the citizenship authorities.

In traditional authorities few tenants' associations are formed, and those that are concentrate on putting pressure on the council to change a policy or improve conditions on the estate. Some of these associations survive for a long time, but others fizzle out quite quickly. Three examples illustrate the key role of the local authority in assisting or obstructing the development of tenants' associations.

Compton

In this citizenship authority Treetops Tenants' Association was formed in response to the discovery of asbestos in one of the blocks. A couple of tenants on the estate contacted the tenants' federation (which was funded by the local authority) for advice on how to establish a tenants' association. The federation contacted the local community worker responsible for developing tenant participation. With advice and assistance from her, the tenants formed a steering group which led to the establishment of a tenants' association.

As the association developed, the members found out about the local authority's system of district housing committees (composed of tenants' representatives and councillors) and started to send a representative to them:

> 'Of course, then we found out why we were never getting any money, we were just getting overlooked every time because no-one was there fighting for money. That's where you put in your bids for your money.'
>
> (Tenant Activist, Compton)

The association became increasingly active, campaigning for the removal of the asbestos and eventually obtaining a charter of rights and compensation for decanted tenants while the asbestos was removed. The tenants' association was involved in joint discussions with the contractors and local authority staff over the improvement and modernisation work to the blocks and, more controversially, became involved in the allocation of three empty flats.

While working closely at times with individual councillors and officers, the association also used a variety of direct action tactics to achieve its goals, on one occasion occupying the members' room in the town hall following a housing committee meeting. During our research, the association was starting to look at the possibility of setting up a tenant management co-operative on the estate, partly as a way of obtaining further resources from central government. As well as the campaigning work of the association, it undertook a variety of social and welfare activities: bingo, a mothers' and toddlers' club, a youth club and an old age pensioners' club. The chair of the association acknowledged the important role played by the council's tenant participation workers who had helped them with advice on how to get furniture and other equipment for their meeting hall.

Treetops developed over time, becoming increasingly sophisticated in its tactics and strategy and maintaining a flexible approach to its landlord. Treetops also illustrates how a citizenship authority continued to listen to and negotiate with a tenants' association which occasionally adopted a highly confrontational approach. This resulted in a basic respect and trust for those councillors and staff with whom the association had contact. So much so that on one occasion when the Department of the Environment arranged to meet the tenants without the local councillors present, the tenants invited two local councillors to come to the meeting in their capacity as tenants. Treetops Tenants' Association also differs from Lowe's evolutionary model in that, from the very beginning, the association combined campaigning with social and welfare activities and many of the activists were involved in both.

Sainsborough

This authority is a large urban authority encompassing relatively affluent and relatively poor areas. Partly as a consequence of this, local politics have been polarised between the Labour and Conservative parties, each winning control of the local authority, sometimes for relatively short periods of time.

During the late 1960s and 1970s, there was a steady growth of tenants' associations coupled with the introduction of formal partic-ipation arrangements. A tenants' consultative committee was formed in 1965 and in the early 1970s non-voting tenant observers were brought into the Housing Committee. By 1975 four representatives were entitled to be co-opted on the Housing Committee with full voting rights and in that year District Management Committees were established with tenant and councillor members.

Officers, councillors and tenants agreed that the arrangement had worked reasonably well and had contributed to strengthening the tenants' movement.

'It was beginning to develop quite well. We had twenty-eight associations and a federation ... I think that's probably what frightened them a bit. It was beginning to pay dividends.'

(Tenant Activist, Sainsborough)

In 1978 control of the council changed hands and the new administration swept away the tenant participation structures. The tenants' associations began to collapse and numbers fell from twenty-eight in 1978 to five in 1988. The explanation of councillors and tenants who were involved was that without any ability to influence the landlord, the associations lost their *raison d'être*:

'It became meaningless. If you can't get anywhere with the council, what's the point of existing.'

(Councillor, Sainsborough)

'Some [tenants' associations] struggled on, but most disappeared – not because the facilities were stopped, but they were actively discouraged. There was no follow-up from the things the tenants' associations were trying to do. They got the feeling they were banging their head against a brick wall.'

(Tenant Activist, Sainsborough)

In 1986 the position changed again and the council adopted a moderately favourable attitude to associations. The tenants' movement was slow to recover, weighed down by a legacy of distrust of the officers who had administered the previous policy of discouraging and ignoring tenants' associations.

Dockport

The Riverside Tenants' Association was formed in 1983 in a mixed development of 1960s-built high-rise and maisonette blocks. One of the main problems confronting the tenants' group was the repair of the estate, which was the reason for their formation. Disrepair had two elements. First, there was substantial water penetration in the high-rise blocks. Second, the maisonette blocks were in a poor state of repair and were targeted for redevelopment by the council.

The council made several unsuccessful attempts to eradicate the water penetration problem. The tenants' association pursued a variety of tactics such as taking the council to the ombudsman (and winning), seeking and gaining extensive press coverage, and actively lobbying councillors. To complicate matters further, asbestos was

140

found in the blocks. The council balloted tenants who supported the line of the association's leadership which was for the asbestos to be left where it was. However, the tenants, perhaps influenced by officers, changed their minds, and window replacement, asbestos removal and limited modernisation were carried out in the blocks. The leader of the association left in protest at the change of heart, putting into doubt the continued existence of the association. Meanwhile, the council put forward plans to demolish and redevelop the maisonette blocks. There was extensive conflict between the tenants' association and the council over the nature of the development plans. There were also disagreements within the tenants' association. Tenants in the maisonettes were concerned that the leadership of the association, who all came from the high-rise blocks, were not adequately representing the views of the maisonette residents and they were critical of the confrontational style being adopted. With some support from the council, who also found the association difficult to deal with, some members in the maisonette blocks broke away from the existing association and formed their own with membership confined to the maisonettes. They fought hard to get the kind of redevelopment they wanted and some of them were closely involved in the design of the new houses. The redevelopment was completed and the new tenants' association members were very satisfied with their new houses and with their influence over the redevelopment. Nevertheless, they were left with feelings of distrust and animosity towards the council. Soon after the completion of the redevelopment the breakaway association folded. Coupled with the uncertainty over the future of the original association, tenant activity in the area has declined substantially.

The history of tenant activity in Riverside shows some of the difficulties which tenants' associations can encounter in a hostile local authority environment. It took the perceived existence of particularly acute problems in the area and a tenant leader with strong, highly critical views of the council to stimulate the formation of the group. Therefore, the origins of the group fitted closely the general proposition put forward by Lowe and others that groups start in reaction to a particular grievance and initially pursue 'agitational activity'. However, this is an oversimplified picture because throughout its existence the group organised social activities for its predominantly elderly membership in the high-rise blocks; and a key influence was internal conflict between two groups of tenants with different interests – an issue rarely addressed in the literature on tenants' associations.

As would be predicted from Lowe's lifecycle approach, the groups changed after the initial grievances were settled and, in the case of the breakaway group, folded. However, this was not brought about by some deterministic process but was the direct result of tenants' attitudes to the council. The tenant activists interviewed thought that there were still plenty of issues to be raised with the council but they and their members no longer felt like fighting.

Activists felt that tenants who attempted to fight the council were singled out, branded as troublemakers and discriminated against. One activist said '. . . I think that they [the council] hold grudges against people that stand up to them, and that is the truth.' Not surprisingly, given these perceptions, tenant leaders felt very vulnerable and isolated and felt that support from other tenants was not always forthcoming, because they themselves felt frightened. In the maisonette blocks in particular there was a high proportion of women who, because of their difficult circumstances as single parents, were thought to feel vulnerable to any reprisals from the council.

In summary, the experience at Riverside would seem at first sight to fit the pattern of origins and lifecycle outlined earlier. However, on closer inspection the pattern was determined not by some innate, impersonal dynamic, but by the actions of the parties involved. In particular, in a situation where the council holds resources, its actions are crucial in shaping the nature and development of tenant activity.

These three examples show that tenants' associations do not follow any determinate, impersonal lifecycle with a dynamic of its own. Instead, their existence and activities are shaped by the actions of their members and of the local authority with which they interact, and the wider context of government policy and the local electorate. A study of individual tenants' associations reveals how their formation and development is contingent on human agency and the attitude of the local authority as well as other contextual factors.

While tenants' individual satisfaction with their housing is related to the quality of the house and service provided by their landlord, dissatisfaction will not necessarily result in the formation and development of tenants' associations. Growth in the number of associations is more likely where the authority adopts a citizenship approach to participation. Then, as in the case of Treetops Tenants' Association, help is at hand from community workers and tenant participation staff with advice and assistance, as well as grants and meeting rooms. Traditional authorities with their lack of interest in, and sometimes perceived hostility to, tenants' associations may frustrate their development by simply ignoring them.

INTERNAL ORGANISATION AND RELATIONSHIP

So far this chapter has examined the motivations of tenants in forming or joining tenants' associations, and the activities of associations. But little has been said about the internal organisation of associations and the experiences and attitudes of ordinary members and more active members. It is to these issues we now turn.

The majority of tenants' associations surveyed had less than two hundred members. A couple had over a thousand members although it is unlikely that many of these were active as in these cases membership followed automatically from a tenancy in the associations' area. Most associations found it difficult to get new members for their committees which implies that active membership is often little more than the committee itself. The majority of committees had between five and nine members.

Most tenants' association executive committees (83 per cent) met at least once a month and about half this proportion had open meetings once a month. The majority of open meetings were attended by at least twenty-five tenants. Nearly one-third only held open meetings once a year. It seems likely that in such cases many tenants on an estate would have had little information about the activities of the association; and in such instances the collective provision of information and consultation by citizenship authorities will be ineffective in reaching the majority of tenants.

It might be assumed that in the citizenship authorities, membership of tenants' associations would be more widespread and tenants' associations more numerous. It was found that there was a relationship between the existence of formal arrangements for participation and the *number* of tenants' associations in an authority, although it was unclear whether the presence of tenant participation structures led to more tenants' associations or vice versa. However, the proportion of tenants who were members of tenants' associations where they existed in their areas was not higher in the citizenship authorities. Membership was higher in the consumerist authority and one of the traditional authorities. In these two authorities, members were more likely than in the citizenship authorities to cite the protection of tenant interests as a reason for joining the tenants' association.

143

Therefore, joining an association may have been seen as a defensive move.

In the previous chapter it was shown that tenants in general had a positive view of tenants' associations and perceived them as useful, even if they would often use other mechanisms such as contacting a councillor. The vast majority of tenants agreed with the statement that tenants should get together on an estate because they had a common interest.

About one-third of tenants who knew of a tenants' association in their area were members. This did not mean that those who were not members were actively opposed to the idea. The most common reasons for not joining a local tenants' association were lack of time, ill-health or no particular interest. In the group discussion held in Sainsborough, tenants commented that they were reluctant to join a tenants' association unless it had an effect on the council. This was in the consumerist authority where tenants' groups had for a long time been excluded from the consultation process.

Joining a tenants' association does not necessarily mean that tenants attend association meetings or are active in any other way. Although tenants are generally positive and supportive of the work of tenants' associations, much of this support is passive rather than active:

'It is just a hard core nucleus of people who are the TA. Although they may have members in number, active people – they're very, very few.'

(Tenant Activist, Dockport)

'As long as somebody is keeping things on an even keel, . . . they (the tenants) really are quite happy with that. They don't want any more than that.'

(Tenant Activist, Compton)

Tenant activists varied in the extent to which they empathised with inactive members. Some lambasted them for their apathy. Others were more understanding of the possible reasons for low attendance at meetings:

'I think lots of them would be interested, but they've got small children and then they can't get out to meetings, you know, because they've got no babysitter or their husband doesn't like them going out in the evenings.'

(Tenant Activist, Sainsborough)

144

Activists were more likely to be out of employment for one reason or another and less likely to have young children than tenants as a whole. Both men and women were active, but they tended to be over 40 years old.

'A lot of the committee are not working. We've got a few retired people on there, we've got a few young mums. Mixture really our committee. We've got middle aged people like myself, we've got a few retired people as I said, and even though T . . . works, he's home at 11 o'clock every morning, so he's available every day as well as I am. D . . . downstairs on our project team, she's a young mum. She has to arrange for a babysitter to come and mind the children while she comes to meetings if it's an afternoon meeting.'

(Tenant Activist, Sainsborough)

As this remark implies, activists often spent a great deal of time on tenants' association work: 'It's getting to the stage where my husband tells me to take my bed over there' (Tenant Activist, Compton).

In authorities which had adopted a citizenship approach to tenant participation, some tenants were spending as much as three evenings a week on tenants' association business. Not only did active tenants' association members often devote a great deal of time to their activities, they were often remarkably persistent. For example, it was not unusual to find examples of tenants campaigning for ten or fifteen years over a particular problem.

Tenants' association office-holders expressed considerable satisfaction with their role despite the work they had to put in and the difficulties faced. Much of the satisfaction seemed to stem from a feeling of having achieved things for other tenants, particularly more vulnerable ones. When asked what she got out of her work with a tenants' association one tenant said:

'A lot of satisfaction, and a feeling that I've been some good to the community. I'm lucky, my two married sons have got their own houses, but then there's young couples here that are never gonna have the chance to own their own home. They got to live in places like X and Y, bring their kids up. So if you can make it a bit better for them. That's really why I enjoy doing it.'

(Tenant Activist, Compton)

145

It was shown in Chapter 3 that many housing managers thought that tenant activists were often out to help themselves, rather than to represent the views of tenants as a whole. However, many tenant activists stressed the role of helping others.

'Very often the person who needs help most finds it may be hard. They would sit and chat as I'm chatting to you now, but put them in front of a telephone and they are petrified and they get tongue tied and you have to be to the point when you're talking about anything to officers of the council.'

(Tenant Activist, Sainsborough)

'To try and make life look a bit more pleasant and hopefully with a bit more community spirit helping one another, either elderly people or even youngsters with families who from time to time need a bit of help.'

(Tenant Activist, Compton)

'I suppose because you think, well your ideals, I mean, you think you'd like to help people, put something back in. I mean I've had problems in the past about getting accommodation for myself and that, and I understand the problems that are involved in it and I do – do feel very sorry for people that are in problems with their housing.'

(Tenant Activist, Compton)

The role of tenants' associations in developing the skills and knowledge of tenants is more frequently mentioned by those working with or promoting tenants' groups, rather than tenants themselves. However, some activists remark on the impact that involvement in a tenants' association has had on the personal development of other tenants, and occasionally themselves:

'By belonging to a tenants' association, I've seen a lot of things I never knew existed and it's made me much more aware of other people's problems. I think it's made me a better person. I feel now that I could go out and cope with a lot of things which I couldn't have done before. For instance, I never could've stood up and spoke in front of people. I would've been too nervous. And I've now joined the school governors and I never would have done that, because I would've thought I wasn't good enough. It's made me realise it's only everyday, ordinary people who do these things.'

(Tenant Activist, Sainsborough)

Although quite a few office-holders had previous experience of the kinds of tasks expected of them, either from work with voluntary organisations or with trades unions or through their jobs, some did not and found it difficult at first.

'I didn't have a clue about anything, and the first time I got up to speak in public I felt really dreadful you know. My knees were knocking, I couldn't get the words out properly. And to take a deputation to the Town Hall and be the spokesman in that Chamber. I'd never been in the Town Hall before in my life. I thought, God, you know, I'm gonna make a real mess of this in a minute. But the second time you go, it's far easier. The third time you go it's nothing. You don't take any notice and you just build your confidence from there.'

(Tenant Activist, Compton)

The satisfactions were enough to keep the activists going to overcome the obstacles. We do not know, however, how many office-holders had given up in frustration or dismay.

'Sometimes I think, oh I'm packing this up. Come to the meetings sometimes at 10 o'clock and my head's banging and I figure, now this is it. I'm giving up. And I say (to my husband) "I'm resigning this year, I'm not going to do it again" and then something happens and you think, no I'll fight on, don't give up, you know, just carry on doing it.'

(Tenant Activist, Compton)

The perceived obstacles related to the lack of support from other tenants and the criticism which was sometimes forthcoming from other tenants.

'You get a certain amount of "We think you're a nosy old devil" you know, "You want to know what's going on" and everything like that. But for every one you get like that I think you get ten who think you've done a marvellous job so it's sort of accept it you know.'

(Tenant Activist, Sainsborough)

One tenant referred to petty vandalism carried out to his car by tenants who thought he was responsible for reporting their behaviour to the council. However, this was very unusual. Most of the difficulties tenant activists face is in their relations with the landlord and it is to this we now turn.

RELATIONSHIPS WITH LANDLORDS

Several surveys of housing authorities have shown a growth in the number of meetings between tenants and authorities since the 1970s (Richardson, 1977; Cairncross, Clapham and Goodlad, 1990; Bines *et al.*, 1993). But such surveys provide little insight into who attends such meetings or into the nature of the relationship between authorities and tenants' associations. Our case studies of six areas provided the opportunity to examine this relationship and to illustrate some differences between and within authorities.

There were great variations within and between authorities in the extent of the contact between tenants' associations and housing management, and between associations and councillors. Broadly, contact was greatest in the citizenship authorities, and least contact took place in the traditional authorities. But there were exceptions to this. For example, two local councillors were frequent visitors to the offices of the only well-established tenants' association in Braehead, whereas the only contact with officers took place through occasional phone calls made by the tenants. In Strathmuir and Compton, in contrast, many tenant activists, and officers as well as councillors, were on first name terms, although they mostly reverted to formal terms of address in public meetings. Officers in these authorities were far more likely than in the other authorities to take for granted that they would spend time at meetings with tenants' associations on a regular basis or by request. For example, in Strathmuir, an annual cycle of meetings to discuss the budget were an established feature of the relationship between the tenants' federation and leading councillors.

Tenant activists' attitudes are affected by factors such as the history of tenant participation in the local authority and its current political composition. The perceived willingness of the local authority to listen to and consult tenants' representatives appears to be as important in shaping their attitudes as the ability of the local authority to provide a high quality service and carry out repairs and improvement work. For ordinary tenants who are not active in tenants' associations, the situation appears reversed. Well-established mechanisms for tenant participation – which they may be unaware of – are less important in shaping their attitude to the landlord than the quality of the service they receive and the ability of the council to repair and improve their homes. Nevertheless, the quality of service was still an important issue for tenant activists, particularly in the citizenship authorities.

> 'The crux of our dissatisfaction with this – with the council is that OK they will listen to us. We have a way of getting through to them, but nothing seems to actually get done.'
>
> (Tenant Activist, Compton)

The question of resources was raised fairly often, and tenant activists perceived an important link between resources and the nature of the relationship with officers and councillors.

> 'The council are not very efficient. There's a lot of things they could do, but when it comes to the crunch line they haven't got the money.'
>
> (Tenant Activist, Compton)

> 'We waited years while they said there was no money.'
>
> (Tenant Activist, Strathmuir)

Tenant activists were generally well aware of the constraints which the council worked under.

> 'I now know that they (the council) have a lot of problems as well, they have to overcome, which you don't really realise when you're a tenant.'
>
> (Tenant Activist, Sainsborough)

> 'We've supported (the council) against the Government . . . We think the (council's capital) debt should be cancelled.'
>
> (Tenant Activist, Strathmuir)

Understanding the point of view of the landlord and the problems which they face can often place tenant activists in a difficult position with their own tenants.

> 'Where people don't appreciate the real problems, you get to defend the council where it's not your job to. In actual fact you know you're cutting the ground from under your own feet by defending the council.'
>
> (Tenant Activist, Compton)

Defending the council against criticism from tenants can be difficult for tenant activists fearful of accusations of having sold out or become 'Uncle Toms'. The problem can be particularly acute in authorities where tenants are given voting powers in committees. For example, a tenant in Compton described the dilemma of being one of a few tenant representatives on a committee which was being asked to decide on a rent increase. The tenants felt they had little influence, but if they had voted against an increase, and won the day, they were threatened that they personally could be surcharged. In this position tenants' representatives were left wondering whether they should represent tenants and suffer the consequences or defend their own interests.

The question of the links between tenant representatives and ordinary tenants was also seen as an important one. Representatives were often unclear who they were to represent and they often received criticism from councillors, housing managers and other tenant activists for allegedly only looking after their own patch. It was unclear what mechanisms of communication existed to tenants outside their own neighbourhood, and no resources were provided specifically for this.

The contact that tenant activists have with their landlord means that they develop a clear idea of the respective roles of councillors and officers and of the power relationship between them. However, their assessments of who makes decisions and, therefore, who they have to influence, varies between local authorities, perhaps reflecting different circumstances. For example, in Sainsborough where there was a hung council, tenants felt that the officers effectively made the decisions. However, in one citizenship authority, Compton, views differed about who was in charge. One tenant activist said that she tried to defend officers when they were criticised by tenants. She told them: 'They (the councillors) make policies, you get them, you have them, don't you have him (the officer). It's nothing to do with him, he's only doing his job' (Tenant Activist, Compton).

However, another tenant felt that behind the scenes it was the officers who were most influential.

> 'I think the officers by and large in Compton are very respon-
> sible people. I think the councillors are, for many reasons, they
> don't match the officers, and the officers have to play second
> fiddle in public, but in private it's the other way round.'
>
> (Tenant Activist, Compton)

In general, tenant activists had a rather sceptical attitude towards councillors. Many tenants praised the support they received from their councillor and said they worked well with them. However, as with tenants in general, respect tended to be confined to councillors who were felt to understand the needs of tenants, perhaps because they were tenants themselves, otherwise their legitimacy in making decisions which affected tenants was questioned. Many councillors were thought to be distant and to look down on tenants.

> 'We have got councillors and you can tell straight away from their attitude, they don't want tenants involved. They don't want tenants telling them what they think of their policies or they made a bad decision. They don't want that at all.'
> (Tenant Activist, Compton)

> 'I think if they have been tenants themselves, even if they've not, then they know people who are tenants, they might know what they go through.'
> (Tenant Activist, Sainsborough)

> 'I don't consider the average elected member is sufficiently qualified professionally. Show me an elected member who lives in the property that's being modernised. If he's a council tenant, yes, take some notice of him. He might have some idea if he lives in the property.'
> (Tenant Activist, Sainsborough)

Tenant attitudes towards housing managers were also mixed. Most built up a close working relationship with their local estate manager and had respect for them. However, the lack of authority and autonomy that these officers had was a constant cause of criticism, and tenants felt that important decisions, for example about resources, were being made by officers they did not have such close contact with.

The question of the authority of the officers with whom tenant activists deal was a recurring theme:

> 'They're there to manage, but they really can't manage as I would understand managing, you know, they always have to refer to somebody else.'
> (Tenant Activist, Sainsborough)

> 'We tried at the bottom and it didn't work. Because as far as I'm concerned, I mean like most of the officers, when we have

a meeting with them, they are only like glorified tea boys. They cannot make any decisions.'

(Tenant Activist, Sainsborough)

For some other tenants who did not have such easy access to staff, relationships were sometimes felt to be more strained and officers were felt to look down on tenants.

'When we first went along we were treated as though we were complete idiots, and they tried . . . the impression I got, you're that big. You haven't got any common sense, you're only a council tenant, so what the hell, you know, and you haven't got any intelligence . . . Now they do treat you with a little bit of respect. But it's been hard work to try and make them realise that we are not sort of just come out of the gutter, because we're council tenants we don't know anything.'

(Tenant Activist, Sainsborough)

In the consumerist authority, staff and officers faced a difficult task in overcoming the legacy of suspicion and distrust left over from the previous policy of ignoring tenants' groups:

'We still have some tenants who regard officers as the enemy and I don't. I see they've got a job to do, we may not always agree with the way they do it, but I'm prepared to argue and discuss with them and generally that's the attitude of most tenants, especially the ones that are involved in any sort of regular interchange with them.'

(Tenant Activist, Sainsborough)

Perhaps not surprisingly, tenant activists in all of the authorities were in favour of tenant participation and all felt that it was in the interests of their landlord as well as the tenants.

'I think in the main they (officers) welcome it (tenant participation) because it must, to a certain extent, make it less difficult for them to find out what is going on.'

(Tenant Activist, Sainsborough)

'Yes, it (tenant participation) helps them to be more effective councillors I'm sure. You know, it's difficult to make some of them see it at times as well. Some of the more sympathetic ones we, we have quite a rapport with and that, you know I think they understand what we're trying to do and are able to assist.'

(Tenant Activist, Sainsborough)

However, there was cynicism expressed by one or two tenants about the motives of officers and councillors in supporting tenant participation, reflecting the self-interest which was thought to lie behind officers' and councillors' actions. In other words, some tenant activists think that councillors and officers are motivated by self-interest, whereas they (tenants) are interested in more altruistic and collective concerns: 'They (the council) are taking notice of what we're saying and they're trying to please us because they don't want the people to go elsewhere' (Tenant Activist, Sainsborough).

CONCLUSION

Tenants' associations form for many reasons and differ greatly in the kind of activities they undertake. Therefore, previous attempts to put forward a universal lifecycle theory of tenants' groups have proved to be misleading. Nevertheless, some pattern can be discerned in the tenants' associations in the different kinds of local authorities identified in this book. In general, in the traditional authorities where the council is at best tolerant and at worst hostile towards associations, few were formed, and those that were, pursued 'agitational activity' to put pressure on the council to improve conditions. Some of these associations survive for a long time, but others fold sometimes in the face of disinterest from the local authority.

In contrast, in the citizenship authorities where support was made available to tenants' associations they were more likely to form for social and welfare reasons or for the development of community spirit. In our consumerist authority, associations had a chequered history as the authority blew hot and cold at different times.

In all areas tenants' associations were generally small, were relatively young and had few resources. Many lacked basic facilities such as somewhere to meet and access to office equipment. Most relied on a small number of activists to keep them going. These activists were motivated by the same factors which led tenants to join associations, namely a desire to improve conditions on their estates. They also felt a basic desire to help other people who were less fortunate than themselves.

Tenant activists derived considerable satisfaction from their role, despite the difficulties they faced. Nevertheless, they were in a difficult position, caught in a kind of limbo between tenants and the council. Their contact with the landlord meant that they were aware of the difficulties and constraints the council encountered and sometimes felt that they had to defend the landlord against what they perceived to be misguided criticism from ordinary tenants. In the citizenship authorities they were also tied into the decision-making machinery. However, even in these cases, tenant activists sometimes felt that they had little influence and that, even where the council listened to them, it was not always possible to achieve the desired outcome. This led to substantial frustration partly because of the distress caused to tenants, but also because their credibility with tenants depended on being able to show results. Therefore, activists saw a danger of being 'incorporated' into the council without achieving much improvement in housing conditions and at the same time being distanced from their own tenants. But, despite the perceived weaknesses in the relationship with their landlord, activists were unanimous in considering that it was worth seeking to build a closer relationship.

In general, tenant activists thought that officers and councillors were supportive and were in favour of tenant participation, although there was cynicism among some activists about why they were supportive. Officers were valued for their technical expertise, but the legitimacy of some councillors was questioned. Those that had lived in council housing were considered to be especially worth listening to, but those that had not were thought to lack legitimacy and to treat tenants as inferior beings. The general stigma attached to being a council tenant was reinforced by the actions and words of a few councillors and officers.

Tenants' associations are more varied in purpose, in their achievements and in the nature of their relationships with councillors and housing managers than the 'agitational' or social welfare roles ascribed to them by Lowe suggests. Tenants feel, in general, that associations have little influence over many issues, but greater influence than individuals acting alone. Whatever the issue, activists keep going because they feel they have made some difference, or will do so in future, even if they do not achieve all they seek. Whether this means that tenants in authorities like Compton and Strathmuir have been incorporated in a way that disempowers rather than empowers them is not an easy question to answer. Ways of seeking an answer are considered in the next chapter.

Part III

THE OUTCOMES

7

TENANTS, HOUSING MANAGERS AND COUNCILLORS

We have concentrated so far on key components of the tripartite relationship between tenants, housing managers and councillors. We have examined the attitudes and objectives of the three parties individually and this chapter aims to provide a framework for examining the relationship between them. Therefore, the focus is on assessing the impact tenant participation has on the relationships between tenants, councillors and housing managers. In particular, what impact does it have on the relative power of the three parties; and does it empower tenants? Power is a key concept in any relationship, and so it is to the literature on power that we look for an analytical framework.

The chapter starts with a review of concepts of power, concentrating on the work of Lukes (1974) and of Clegg (1989). Clegg's view of power as a complex game made up of players engaged in shifting and unstable alliances is taken as the framework for the analysis. The chapter continues by examining the tenant participation 'game', focusing particularly on the role of tenants as individuals and as part of a collective such as a tenants' association. The processes by which the 'rules of the game' are decided are then examined. The rules may be influenced by the results of the game between the three parties or they may be determined by external factors such as central government policy or legislation. The influence of these factors is shown by a comparison of tenant participation in the 1970s and the 1990s. Finally, conclusions are drawn about the nature of the tripartite relationship in tenant participation.

TENANT PARTICIPATION AND POWER

The most widely used approach to the concept of power since the middle of the 1970s has been that of Lukes (1974). He developed a three-dimensional view of power which sought to build on insights derived by Dahl (1961) and Bachrach and Baratz (1963) on decisions and non-decisions. Lukes included these two approaches as his first two dimensions and added his own third dimension emphasising the structural influences on social definitions of meaning and knowledge:

> the three-dimensional view incorporates power of the first two kinds, but also allows that power may operate to shape and modify desires and beliefs in a manner contrary to people's interests. In consequence, neither revealed preferences, nor grievances and inchoate demands will always express them ... Power, on this view, may encourage and sustain attitudes and expectations that work against people's 'welfare interests' or subvert and thwart their pursuit of their ulterior, focal aims or both.
>
> (Lukes, 1986, p.10)

The implication of this approach is that tenants (or any other social actor) may fail to identify their objective interests because of the power of other agencies to influence people's perceptions of the world.

The major problem with the 'third dimension' is in the definition of the 'real' or 'objective' interest. Lukes realises that this may be defined in different ways by different people depending on their particular standpoint, whether they are a liberal, a reformist or a radical. Although Lukes adopts a radical position, he fails to put forward a model able to reveal what 'real' interests are from this standpoint. Therefore, it seems to be open to individual commentators to reach their own conclusions, leaving the way clear for many different interpretations.

Lukes' third dimension of power presents great difficulty to any researchers who wish to detect it on the ground. For example, what rent level is in the real interests of tenants? One's first inclination may be to reply that the lower the rent, the better for tenants. But if the price of low rents is a bad service and inefficiencies generated by underfunding, is it possible to assert wholeheartedly that low rents are in the tenants' objective interests? In the case of repairs, it is not clear that the right of an individual tenant to get compensation for delays in carrying out repairs and an order for their immediate completion, is in the objective interests of all tenants if it results in the diversion of funds from an already depleted maintenance and repairs budget, with consequent delays in planned repairs and maintenance programmes.

Clegg (1989) criticises the relativism of Lukes' third dimension and builds an alternative model on a different historical tradition of thought. He places Lukes in a tradition of writers on power that goes back to Hobbes. This school of thought is one that is concerned with what power *is*. In this tradition, power is seen as:

a locus of will, as a supreme agency to which others would bend, as prohibitory; the classic conception of power as zero-sum; in short, power as the negation of the power of others.

(Clegg, 1989, p.4)

Clegg argues that there is another tradition which predates Hobbes, originating with Machiavelli, that is preoccupied with what power *does*.

Power is simply the effectiveness of strategies for achieving for oneself a greater scope for action than for others implicated by one's strategies. Power is not any thing nor is it necessarily inherent in any one; it is a tenuously produced and reproduced effect which is contingent upon the strategic competencies and skills of actors who would be powerful.

(Clegg, 1989, pp.32–3)

Against the Hobbesian tradition with its causal, agency approach to power, Clegg refuses to accept any single originating and decisive centre of power and describes a world of shifting and unstable alliances in a state of flux: 'power is best approached through a view of more or less complex organised agents engaged in more or less complex, organised games' (Clegg, 1989, p.20).

For Clegg, the key to understanding resides in:

thinking of power as a phenomenon which can be grasped only relationally. It is not a thing, nor is it something that people can have in a proprietorial sense. They 'possess' power only insofar as they are relationally constituted as doing so. To the extent that the relational conditions which constitute power are reproduced through fixing their obligatory passage points, then possession may be fixed and 'reified' in form.

(Clegg, 1989, p.207)

The concept of 'obligatory passage points' is a key one in Clegg's framework as it represents the arena in which battles are fought or, to use his analogy, it constitutes the 'rules of the game'. Therefore, 'obligatory passage points' are the conduit through which power relationships are expressed. However, these passage points may be fixed only temporarily and may be changed by exogenous factors or through the game itself. This is because reification or fixing occurs rarely without resistance, which may take two forms. The first and most common is resistance to the exercise of power within the existing rules of the game. An example may be tenants and their landlord seeking to work out the details of a modernisation scheme through participation mechanisms, with tenants seeking to influence issues which councillors or officers have previously seen as their domain.

The second form of resistance is what Clegg calls 'organisational outflanking' where resistance to power may consolidate itself as a new power with a new relational field of force (1989, p.207). In other words the 'rules of the game' are themselves contested and may be changed creating a different 'obligatory passage point'. This may be done by contesting the meaning in which one is being implicated by others. For example, tenants fairly often dispute their landlord's definition of tenant participation as information provision, arguing for a new meaning to be accepted and formally agreed, where tenant participation means consultation or negotiation.

Clegg's emphasis is on the fluid and contingent nature of power. He directs attention to the arenas of power play, to question why *this* arena and not *that* one; to look at the rules of the game and the way it is played or organised and the skills and strategies of the actors. He emphasises the role of organisation in marshalling the resources available to actors and the strategies which they pursue, including the creation of networks and alliances. Clegg also stresses the 'relational' quality of power. In other words, power exists and is manifested only in the relationships between different actors. The individual in isolation is neither powerful, nor powerless, as power in such a context is meaningless.

Clegg's framework is a useful one for analysing tenant participation. The structures and processes of participation can be viewed as 'obligatory passage points'. Attention can then be focused both on the game which takes place within these rules and also on the influence of the game and of exogenous factors on the 'fixing' of the rules, in other words, the nature of tenant participation. In viewing tenant participation through Clegg's lens, the nature of tenant participation mechanisms – the 'obligatory passage points' – may be seen as the structures, processes and objectives of tenant participation identified in Chapter 2. Therefore, the three models of traditionalism, consumerism and citizenship may be seen as different rules within which tenants and their landlord play the participation game.

PLAYING THE GAME

Tenants can participate in housing management as individuals or part of a collective such as a tenants' association. In practice, both forms of participation take place in each of the three types of local authorities identified in this book. Thus even in the individualist, consumerist, authority tenants may join together to try to exert influence on their landlord, despite the reluctance with which some councillors respond to them. In the collectivist, citizenship authorities, tenants clearly still have an individual relationship with their landlord as well as possibly a collective one through a tenants' association. One of the reasons for the coexistence of individualist and collectivist tenant involvement is the ambiguity of tenant attitudes on this issue, as shown in Chapter 5.

Two major factors influence the game the parties play within the established rules. These are organisation and tactics. Organisation is used here in the broad sense adopted by Clegg (1989) meaning the effective utilisation of resources at the party's disposal. The concept of resources is used to cover all the different kinds of capacity which go to make up what Clegg calls dispositional power. This section will start by looking at the individual tenants' involvement in the game before examining their collective involvement. The focus is on organisation and tactics in each of these situations.

INDIVIDUAL TENANTS

Consumerism has focused on four issues which are of relevance to individual tenants: access; choice; information; and redress. Each of these will now be considered.

The six local authorities considered here had all taken action of some kind to improve the access of tenants to services. The most general move was to improve physical access by decentralising offices and making them more welcoming. The training of frontline staff was also given some attention particularly in the consumerist and citizenship authorities. The traditional authorities had not adopted these changes to the same extent.

161

Individual tenants in all the local authorities were given opportunities to exercise choice. For example, in most of the authorities tenants could choose the colour of their bathroom suite when their house was modernised. In two authorities (one traditional and one citizenship) individual tenants also had some choice over what modernisation funds were used for. In other words they could choose whether to have a new bathroom suite or cupboards in the kitchen.

The provision of information to individual tenants was undertaken in all local authorities. All six sent letters to individual tenants and produced leaflets, handbooks and information packs. Again, the provision of information was more extensive in the two citizenship authorities and to a lesser extent in the consumerist authority. Information provision was often well developed on issues such as modernisation where choice was available. In some cases, again particularly in the citizenship authorities, independent agencies were available to provide information and advice to tenants about the services and how to register a complaint.

Redress mechanisms have in the past been informal, largely consisting of contact with housing staff or with a local councillor. This remained the case in our traditionalist and consumerist local authorities. In addition to local procedures, tenants can take some complaints to the ombudsman when all other avenues have failed; and a number of tenants in the six authorities had taken this route. However, the decisions of the ombudsman are not binding on councils.

In the two citizenship authorities formal systems of redress through grievance procedures existed. In Strathmuir, for example, the tenancy agreement resulting from a process of negotiation between tenants' representatives and the council in the early 1980s included provision for a tribunal to consider disputes or complaints arising between the council and tenants. It took many months for the council to agree the regulations under which the tribunal would operate, and months more before the membership was put in place. Each appeal was to be heard by a tribunal of three people, one councillor (appointed from a panel), one tenant (from a panel nominated by the tenants' federation) and one independent person. Only careful reading of the tenancy agreement, or contact with councillors, officers or others familiar with the system would have led any tenants to use the tribunal until the early 1990s – ten years after it was devised. The council was hesitant about advertising the system widely in case it was overwhelmed by the response. Some publicity was generated after pressure from the tenant and independent panel members of the tribunal. As a consequence, only a handful of hearings had taken place during the period of our research. Although the regulations allow the tribunal to consider most issues that arise in relation to the terms of the tenancy agreement, the limitations of the agreement itself make it unlikely that the council will be faced with, for example, large repair obligations arising from precedents established at the tribunal.

In using opportunities and mechanisms open to them the most important resource of individual tenants is their own knowledge and skill. For example, knowledge is necessary to be aware of the opportunity to exercise choice and to make the most appropriate choices. As an individual consumer, a tenant may be unable to exercise this choice either because they cannot gain access to other landlords' stock, or there are no alternative landlords available or because they cannot afford to buy. In addition tenants need knowledge to be able to take advantage of any alternatives.

In the participation game, tactics are also important. Knowing who to contact, at what stage and how to approach them can be crucial in determining the outcome. In Chapter 5 we showed that many tenants would enlist a 'prominent person' such as a councillor or MP to help them in any dispute with the landlord. This came from a perception that such people had more clout and perhaps could put the case more effectively.

At the end of the day tenants as individual consumers have few resources at their disposal. Tenant's Choice has been very little used, and tenants' actions suggest they do not consider it a liberating right. Some tenants may be able to take advantage of the Right to Buy their house but many others will not. Many cannot choose to take their custom elsewhere. Some local authorities have allowed tenants extensive control over the use of their home by allowing choice in modernisation schemes. However, the scope for effective individual action is limited both by the rights and duties of the landlord as owner of the housing and the collective nature of housing as a service. Some local authorities may listen to individual tenants and act accordingly but, if it is not in the interests of the landlord to act or if they choose not to act, tenants have few resources to deploy. Tenants have only limited legal rights and usually only the ombudsman as a very limited form of independent redress. The sanctions that an individual tenant can exert over local councillors are small in that they can probably sway only a few votes.

In the consumerist and the citizenship local authorities considerable effort was expended in improving choice for individual tenants, in the provision of information, and in creating grievance procedures and forms of redress. However, some tenants were still motivated to join together in order to increase their resources. As we argued in Chapter 5, individual and collective action were seen as complements rather than alternatives and most tenants were happy to pursue any route they thought would be effective in a given situation. It is to the collective mechanisms that we now turn.

COLLECTIVE INVOLVEMENT IN THE GAME

Tenants' associations have a number of resources at their disposal in playing the housing management game. Probably the most important of these is the support of the tenants as a whole. We showed in Chapter 5 that most tenants expressed general support for tenants' associations even though mostly passive.

Many tenants perceived a common interest with other tenants, but the extent of this perception varied considerably between tenants in different local authority areas and in different estates in the same local authority (Cairncross, Clapham and Goodlad, 1993). In addition, tenants' views on the problems facing them, and their perceptions of their landlord also varied considerably. Therefore, council tenants do not form a unified social base and do not face a common consumption position as argued by Saunders (1979), Dunleavy (1980) and Lowe (1986). (For a fuller discussion of this argument see Cairncross, Clapham and Goodlad, 1993.) Mobilising support from tenants is one of the most important tasks facing tenant representatives. The achievement of this task may depend on the means of communication with tenants. However, it was shown in Chapter 6 that many associations lack basic facilities such as office space or a photocopier and do not have the income to acquire them.

Other resources which tenants' associations need to organise are skills and knowledge. Tenants' representatives need to be able to manage the activities of the association and be able to communicate with other players in the game. Some tenant representatives may have these skills from their employment or from previous activities. For example, some may have supervisory jobs or be trade union representatives. Some skills training is available for tenant representatives as well as conferences and courses which could be used to improve knowledge of housing management and other important issues (for a review of training see Scott *et al.*, 1994). Tenants' associations have little income to be able to purchase this training. Although in Compton and Strathmuir the councils had provided substantial grants to tenants' associations and their federations, in comparison with the resources housing managers and councillors command, even the richest tenants' federation was poorly resourced. Tactics are also an important part of the game. As Clegg argues:

> Frequently those who are relatively powerless remain so because they are ignorant of the ways of power: ignorant, that is, of matters of strategy, such as assessing the resources of the antagonist, of routine procedures, rules, agenda setting, access, of informal conduits as well as formal protocols, of the style and substance of power.
>
> (Clegg, 1989, p.221)

Tactics may vary considerably depending on the 'rules of the game'. Where an exchange of information or a dialogue takes place as in the citizenship authorities or in difficult-to-let estates in the traditional authorities, tenants are expected to adopt 'professional' tactics. In other words they are expected by the landlord to state their position clearly, to produce a logical and well-researched argument to back up their case and to engage in constructive and even-tempered dialogue.

> 'Never lose your temper. Rule One. A well-reasoned sensible argument verbally given is far better than a tirade . . . You've got to keep calm and collected to argue a point through very clearly, I think.'
>
> (Councillor)

> 'Well I would advise them first of all to do a survey on the estate.'
>
> (Housing Officer)

Housing managers and councillors were critical of tenants' representatives who adopted a hostile attitude at the outset or who resorted to other tactics such as using the press or holding demonstrations before the landlord had a chance to engage in dialogue. This kind of tactic was said by housing managers and councillors to make relationships more difficult and to be a less effective way for tenants to achieve their aims. However, it was regarded as legitimate for tenants' representatives to indulge in these kinds of tactics if constructive dialogue had failed. In addition, in local authorities where there is resistance to a dialogue, as in some traditional and consumerist authorities, tenants' representatives may feel they have to use such tactics to get a dialogue going or in place of a dialogue.

Tenants' representatives may not meet resistance if the landlord agrees with their point of view, but often some resistance will be encountered. In these circumstances tenants' representatives have to attempt to employ what Clegg (1989) calls 'agency' power, that is one agent causing another agent to do something she otherwise would not do. The force of resistance will depend on the organisation and tactics of housing managers and councillors. Councillors have the resources of democratic legitimacy and may be able to draw on support from other councillors, particularly from those in their own party. Also, councillors are able to use the resources of the council, including the skills and knowledge of staff and the council's financial resources. These may seem to be large but, as with tenants, resources need to be organised. Support from other councillors cannot always be taken for granted. A 'power game' may take place within the party group in which councillors have to organise their own resources and in which tactics may be important. Also, the democratic legitimacy of councillors may not last for ever. For example, councillors in marginal wards or a party facing an electoral threat to its control may have fewer resources to draw on than councillors in safe wards or a party encountering little electoral opposition. Also, some councillors or parties may depend on support from tenants which could be threatened by a well-organised tenant campaign whereas other councillors or parties may not: 'They'd [councillors] go along with anything at this moment in time because all they want to do is justify themselves, think of the next election' (Tenants' Association Member).

The resources of the council also need to be organised through the exercise of control over the organisational functioning of the council. Individual councillors and parties may vary considerably in their relationship with council officers and in their ability to exercise control.

Housing managers are the third party in the game, who may have their own objectives, resources and tactics. Their major resources are their professional skills and knowledge and their access to council resources. These also need organisation. Professional skills and knowledge have to be deployed and their legitimacy accepted by

166

other parties. Access to council resources may be challenged by councillors. A favourite tactic we encountered in the case studies was the attempted forging of alliances between the parties. Thus, housing managers would seek the support of tenants in attempting to influence the councillors and so on.

One symptom of tenants' powerlessness was demonstrated in their participation in local political arenas. For example, in Compton there were long delays in getting officer reports on matters raised by tenants, so that items were repeatedly carried over to future meetings. In Strathmuir, issues raised by tenants were sometimes unresolved in discussion and effectively dropped from the agenda. A common device, especially in traditional authorities, was to question the legitimacy of those tenants' groups which took a confrontational approach to the landlord, or tried to introduce issues onto the local political agenda which the local authority wished to exclude. Whether or not these were deliberate tactics by the landlord is unclear, although they were frequently perceived as such by the tenants affected.

The resources which the players bring to the game not only depend on their own actions but also on the framework within which the players operate. For example, the willingness of councils to listen to tenants has been increased in the 1980s and 1990s by central government policy. The introduction of the Right to Buy in 1980 and the 'Tenant's Choice' clauses of the 1988 Housing Acts have provided tenants with the option of either owner-occupation or having a new landlord, and the provisions of the Leasehold Reform, Housing and Urban Development Act 1993 give tenants the right to take over the management of their housing from the council. Tenants may now use the threat of exit from local authority tenancy or management as a device to obtain benefits from their landlord. Council landlords were very conscious of the significance of the 1988 legislation and talked about the importance of improving the quality of service provided to tenants and developing tenant participation as a safeguard against a large scale exodus. Housing managers and councillors conceded a change in the rules of the game, irrespective of the attitudes of tenants.

'Tenant's Choice and competition generally has made author-ities sit up and think well, it is no good any more accepting that the local councillors know what the people want . . . .'

(Chief Housing Officer)

'Because of the Housing Bill, we have to have more tenant participation. We have to look at tenant management co-operatives, we have to work closer with tenants if we want to keep the properties. If we want to make sure there's security for tenants, then we have to work closer with them to keep that security. We have to put the tenants first.'

(Councillor)

In practice, although tenants have not rushed to take up Tenant's Choice, a growing number have shown interest in tenant management (Scott et al., 1994). However, with the passage of time, the impact of Tenant's Choice may have lessened, and in 1996 it was scrapped by the government.

Even when tenants and landlords have negotiated an agreement, the tenants may find that external forces have influenced the outcome. This is shown in the examples of Compton and Strathmuir where tenants held apparently successful campaigns to have a say in drafting new tenancy agreements, following the 1980 housing legislation. Although the councils were initially perceived as reluctant to give tenants any input, the tenants were able to achieve many of their aims in the agreements that were eventually negotiated.

One of these agreements in Compton has been widely considered a model of its kind in the rights it gave to tenants. By 1989, however, it had become the subject of ridicule among tenants, councillors and housing managers. The state of the housing stock, the lack of resources, staff shortages in the housing department and a variety of other factors combined to make the tenants' rights enshrined in the agreement unenforceable, and the agreement itself more a statement of intent than a description of current practice: 'I think it's [the tenancy agreement] wonderful, I only wish we had the resources to match it' (Director of Housing).

In Strathmuir a similar, if less extreme situation developed with budgetary constraints preventing certain repairs being carried out as had been done in the past. While not breaking the terms of the tenancy agreement, this development dismayed the tenants' federation which had reluctantly accepted a weak repair obligation on the council at the time the agreement was negotiated. Therefore, the result of the game may be strongly influenced by factors exogenous to the game itself, and these factors have intensified since 1989 with the imposition of a new financial framework for revenue accounts in England and Wales.

Another example illustrating the nature of the tenant participation game and the numerous factors influencing the outcome is the campaign by tenants in Compton after the discovery of structural faults in their estate. Their campaign began not long after a by-election defeat in the area, which had been regarded as solidly Labour. With assistance from a local law centre, the tenants were able to negotiate a charter from the council, guaranteeing them the right to be rehoused (on completion of major works to the estate) in properties with one bedroom surplus to the authority's definition of housing need. Working with the tenants' association, the council was able to secure permission from central government for a capital improvement programme which would also include some new houses.

Over time, the decision to grant the Tenants' Charter and to allocate the new housing to tenants from the estate has been challenged. Critics have argued that the estate was in a better state of repair than others in the authority; that the planned allocations were implicitly racist because most of the tenants of the estate were white (reflecting institutional racism in allocations in the past) and this did not reflect current housing need among black and white people across the authority. It emerged that this was probably the last new-build scheme in the authority for some time, because of the cuts in the authority's capital programme. A local black community group took the matter to the Commission for Racial Equality (CRE) and at the time fieldwork stopped, the houses were standing empty awaiting the CRE's decision. Meanwhile the council had started letting empty properties on the estate to homeless people, reversing its earlier decision to cease letting.

Over the years the original fragile alliance between tenants, council and central government broke up. The alliance between the council and the tenants had developed in the wake of the by-election. Changes in the composition of the council, the development of the black community group and the CRE created a new dynamic and a new set of rules.

DECIDING THE RULES OF THE GAME

The playing of the game can result in a particular decision being made, but it can also result in the rules of the game themselves being changed. An important example of this concerns attempts to control the agenda or the issues to be addressed through participation.

Most tenant participation is on 'safe issues' – as demonstrated by the result of the national survey of local authorities. Over 80 per cent of local authorities in the survey consulted tenants on modernisation and rehabilitation and over 60 per cent on estate management. One-third or fewer consulted tenants on issues such as repairs and allocation policies, and just over one in ten on rent levels. Of the tenants' associations in the six case studies, one in five were directly involved in trying to influence the local authority on rent levels; more than one-third on the tenancy agreement; and two-fifths on allocations. Thus, although at least four-fifths of British local authorities have provision for some kind of tenant participation, tenants appear to have been relatively unsuccessful in putting their grievances onto the local political agenda.

Most tenant participation has been confined to local management matters, rather than wider policy issues. In the case studies, attempts by tenants to broaden the participation agenda beyond such issues were often criticised by housing officers and councillors as outside the legitimate realm of tenant activity. For example, attempts in Sainsborough by tenants' groups to question the subsidy of the General Rate Fund by the Housing Revenue Account were blocked by staff and Conservative councillors. It was seen as an issue inappropriate for tenant participation, although it had clear consequences for rent levels.

'They [a local tenants' association] go into policy matters rather than just say "The street lighting isn't working down the end of the road." It's "Why is the Council doing this? Why is the HRA funding the Rate Account?" And you think, is this a tenants' association meeting ... We don't want to put them down as it were, but it's very little to do with local issues and again you wonder if it's representative of those residents and would they like to know that that's what their committee talks about when they come to see us, is funding and financial arrangements.'

(District Housing Manager, Sainsborough)

Sainsborough also illustrates how changes in political control can have a profound influence on the 'rules of the game'. For a period in the early 1970s the council was controlled by the Labour Party and a well-developed structure of tenant participation was in place, supported by tenants and considered to be successful in allowing them to put issues on the agenda.

'I think we were listened to much more and obviously the opportunity was there for putting in opinions and suggestions and so on directly right into their own committee and the housing management committee as well.'

(Tenant Activist, Sainsborough)

In the local elections of 1978 the Conservative Party took control and the tenant participation structure was dismantled and tenants' associations were ignored.

'(Since 1978) some (tenants' associations) have struggled on, but most disappeared – not because the facilities were stopped, but they were actively discouraged. There was no follow-up from the things the tenants' associations were trying to do. They got the feeling they were banging their head against a brick wall.'

(Tenant Activist, Sainsborough)

It was during the Conservative administration that consumerist practices emerged and almost all contact was with individual tenants rather than with tenants' associations. In 1986 the Conservative administration lost power and the council became 'hung' with no party in overall control. Also, no formal arrangement for the sharing of power between the parties was agreed. This change in political control opened the floodgates to a steady stream of issues which could be debated for the first time in years in the housing committee. Tenants were able to lobby senior officers and councillors to put their grievances on the agenda. Previously the tenants were unable to move the issue from covert grievance to observable conflict. In other authorities, where there was no authority-wide forum for tenant participation, tenants could not begin to get such grievances discussed (should they have wished to). In contrast, in Compton and Strathmuir, the councillors took it for granted that rent levels and subsidies were appropriate topics for dialogue – their rules of the game differed from councillors in the other authorities.

Another illustration of an alteration of the rules comes from Compton which had a long history of tenant participation. The first formal structure for tenant participation was a tenants' consultative committee, which was composed of councillors and an equal number of tenants elected from a tenants' organisation covering several different authorities. New arrivals to the local tenants' movement saw the committee as an ineffective and council-led body. Their main criticism was that the consultative committee monopolised

171

access to the council and that membership of the committee depended on affiliation to the tenants' organisation.

'I contacted the [tenants' organisation]. After talking to them, it sounded like one of the most undemocratic groups ever. They said "You have to join our organisation. We're the only group the council will talk to."'

(Former Tenants' Association Chair)

The relationships of power had been reified in the form of the tenants' consultative committee and this was the channel through which power flowed. The tenants on the consultative committee were mostly from one geographical area within the local authority with strong links to the local Labour Party. The campaign for a new district committee-based structure for tenant participation was led by tenants from another part of the authority, less closely associated with the Labour Party. Their eventual success was ascribed by those involved to a variety of factors which included a major change in personnel on the council following local elections. New councillors, some with backgrounds in the tenants' movement, a new Chair of Housing for the first time in more than fifteen years, and a new officer responsible for promoting tenant participation, combined with the influx of a large number of active and experienced tenants' associations following a stock transfer to the authority, created a new set of alliances. The relationships continued to evolve as plans for further decentralisation developed which may widen involvement beyond tenants' groups to other community groups in each neighbourhood. Thus, at each point, when power became 'reified' as Clegg puts it, groups have resisted the exercise of this power and new groups and alliances have emerged.

The rules of the game are often structured to offer a tactical advantage to councillors and housing officers by dividing tenants and obliging them to compete with each other. For example, in Compton, the participation arrangements were structured in such a way that tenants' associations had an incentive to join formal participation arrangements to gain access to the district committee where decisions about estate improvements were made. Tenants' associations concentrated on playing the game and competing with each other rather than changing the rules.

'We had nobody going onto district committee. We had no-one kicking up any kind of fuss, [saying] "Hey, look what's wrong with our estate, with our dwellings and everything like this." So then we found out about that and put one of our reps, or vice chair, onto the district committee. Because we found out why we weren't getting any money. We were just getting overlooked every time someone's there fighting for money.'

(Tenant Activist, Compton)

Once access had been gained tenants' associations had to compete against each other to get improvements to their estate.

'I think the thing I find unfair about that, and it is a personal opinion, is that the person who shouts loudest, is going to get the best or the most. I think it's the old, old story – if you've got a strong T.A., in turn a strong representative on the district committee, then the estate is going to get what's going or a fair big chunk of what's going . . . whereas if you've got a nice, genteel tenants' committee who don't shout very loud, but would very much like a lot of money, they might have more problems, but if they haven't got somebody with a loud voice or someone who knuckles in, they may not get what is due to them, or what they deserve.'

(Capital Programmes Officer, Compton)

Competition between estates and local authority areas intensified with the development of the government's Estate Action programme, and later the 'Single Regeneration Budget'. Here councils have to bid for loan sanction to enable them to pursue physical regeneration of an estate. One of the criteria for approval is that tenants are consulted and support proposed changes. Therefore, central government has provided the stimulus for the initiation of tenant participation on some estates.

Exogenous factors such as central government policy can be very important and lead to changes in the 'rules of the game'. For example, following the 1980 Housing Act, councils have been obliged to consult tenants over changes in housing management which substantially affect them. This has encouraged participation. However, some issues, most notably the level of rents, have been explicitly excluded from these provisions, and this has influenced the rules of the game in some authorities and made it difficult for tenants' groups to get this issue on the agenda.

The exogenous factors which influence the 'rules of the game' are much wider than central government policy. It is impossible to outline all of the factors which could possibly be important. However, some insight can be gained by contrasting the different circumstances surrounding tenant participation in the mid-1970s and the mid-1990s.

During the late 1960s and early 1970s there was considerable tenant activity around the issue of increasing rents. As Hague (1990) reports, in 1968 the Labour government sought to increase council rents and introduce a national means-tested rebate scheme in the wake of the devaluation of sterling. Then the Conservative government's Housing Finance Act of 1972 sought to push up council rents to the 'fair rent' levels set in the private sector. Both measures led to sustained tenant campaigns which, although they were ultimately defeated in their major aim of preventing rent increases, did result in the introduction of tenant participation arrangements and changed the 'rules of the game' by legitimising tenant activity.

The tenant campaign has been described by Lowe (1986). At times it was a national campaign and it included a number of local rent strikes and a national non-compliance campaign by some local authorities. The extent of the campaign led Lowe (1986) to characterise the tenants' movement as an urban social movement.

The important question is how the tenants' movement was able to alter the participation 'rules of the game'. Richardson (1983) argues that the active tenants' movement was just one part of a general growth of consumer demands. She argues that people had rising expectations of services, including public housing. At the same time, she argues that there was a change in public attitudes towards authority and an increased willingness by people to make their views known to those in authority and to demand that services meet their needs and provide value for money. In other words she argues that there was a cultural change in the late 1960s which enabled tenants' associations to mobilise their members.

Richardson also argues that service providers such as housing departments faced problems caused by the increasing growth and scale of local government services. Local government was reorganised into larger authorities during this period, and the rapid growth of council housing in the 1960s had increased the scale of housing management. Tenant participation could give service providers information on which to base decisions and help ease the housing management problems which were emerging in the early 1970s. Thus, she argues that council landlords were more open to tenant influence in order to help solve their own problems.

Hague (1990) sees tenant participation at this time as a response to a financial crisis in the economy. Therefore, the Labour government in the late 1960s was pushed into action by problems with the economy which forced them to reduce public expenditure and attempt to withdraw housing subsidy. However, the government was in a weak political position and needed to retain the support of tenants. Labour local authorities too were weak politically at this time as they shared the government's unpopularity. Many of them, too, depended on tenants' support.

During the early 1970s there was a Conservative government which suffered from economic problems but which did not need tenant support as the Labour government had done. However, opposition forces attempted direct action through trades unions and the tenants' movement to counter the free-market policies of the government, and this paid off when the Conservatives were later voted out of office following a coalminers' strike.

Thus, in the late 1960s and early 1970s tenants had a common issue (increasing rents) which involved all tenants. There was a climate of increasing questioning of authority and a context of direct collective action in a number of fields. Central governments either depended on tenants' political support or were vulnerable to campaigns in a difficult economic situation; and local authorities were either vulnerable to political pressure or supported tenants in order to challenge central government or to assist them with housing management problems. In such a situation tenants were relatively powerful in relation to landlords and the 'rules of the game' were changed, with arrangements for tenant participation spreading and landlords engaging in a dialogue with tenants.

The position by the early 1990s was very different. Council housing had declined in absolute size and had become largely an urban phenomenon as 'Right to Buy' sales are greater in rural areas. The sector contained an increasing proportion of disadvantaged households as most of those able to enter owner-occupation had done so. The outcome was the declining political power of council tenants at central government level. Of course, council tenants remained an important force in some urban local authorities where large stocks of council housing remained. However, the decentralisation of housing management and the emphasis on action on individual estates made it difficult for tenants to perceive a common interest (Cairncross, Clapham and Goodlad, 1993). The defeat of the tenants' movement on the rent question in the early 1970s meant that, since that date, the question of rent levels has largely been removed from the agenda. Since then the introduction of housing benefit has meant that over two-thirds of tenants do not pay their rent in full and so are insulated from rent increases. This also means that the rent strike is a less valuable weapon than in the 1960s.

There have also been cultural changes. During the 1980s and 1990s strong Conservative governments pushed the cultural values of individualism and of market relations. In housing this emphasis is illustrated by the Right to Buy, by compulsory competitive tendering in housing management, and by moves towards consumerism as demonstrated by the Citizen's Charter and by the consumerist approach to tenant participation outlined earlier. Collective action, as illustrated for example by trades union activity, has declined considerably since the early 1970s, and has not been entirely replaced by activism around other issues such as conservation, race and gender.

Nevertheless, the introduction of the Right to Buy, Tenant's Choice, the right to manage and compulsory competitive tendering have given tenants a way out of council housing which threatens the size and authority of local authority housing departments and the jobs of their staff. Therefore, the incentive has been created for council landlords to engage in tenant participation in order to be able to respond to the wishes of tenants. In addition, central government support for tenant participation through legislation and promotion have been important in setting the 'rules of the game'.

These general factors have influenced the growth of tenant participation and the form the 'rules of the game' have taken. However, the variety of practice between local authorities is wide. In the early 1970s support for the campaign against rent rises, and the form that protest took, varied considerably. In the 1990s the variety in extent and form of tenant participation is shown by the differences between the traditionalist, consumerist, and citizenship models outlined earlier. Variations in response to exogenous factors exist because of the differential impact they have on the local power relationship between landlords and their tenants.

CONCLUSION

In this chapter we have examined the impacts of tenant participation on the relationship between the three parties involved. The approach to power advocated by Clegg (1989) has been used to explore the nature of tenant power. The use of this approach shows the contingent and fluctuating nature of tenant power. The way that tenants and housing managers and councillors play the game, the tactics they use and the resources they can deploy all influence the result. However, many other factors outside the control of the immediate actors, such as the political culture and the position of council housing are also important, both in influencing the outcome of the game and the rules by which it is played.

Tenants' power is a fragmented and shifting phenomenon which exists only in relationship to other parties. In the local authorities in our research it was the process of tenant participation and the relationships between the actors that emerged as important rather than the formal arrangements. When tenants appeared to have 'won', it was not always a permanent victory. Changing relationships, new alliances and altered circumstances could and did undermine these seeming victories. Also victory of one group of tenants was often at the expense of another group. We have discussed the wide variety of factors on which tenants' power may be contingent, and the way that it may change over time. On their own, structures for tenant participation cannot increase tenants' power unless other factors such as the implicit or explicit threat of sanctions combine to operate in its favour. And if these factors change, alliances break up, or new groups emerge, tenants' power may just as quickly go into decline. As we have shown, the relationship between council tenants and their landlords has in many ways altered to the advantage of tenants in the last twenty years. But this conclusion must be qualified with reference to the local and exogenous factors which influence the relationship to provide more skills, opportunities and sanctions to landlords, even when such resources are reduced relative to central government.

The contingent and fluctuating nature of the power relationships in tenant participation makes it difficult to draw firm conclusions about the relative merits of the three local authority models. Clearly the citizenship model in theory offers tenants the opportunity to exercise individual and collective influence. Individual influence is encouraged through the provision of information, the maximisation of possibilities for choice and through surveys and the creation of grievance and redress procedures. Consumerist authorities may do all this, but citizenship authorities add structures and processes of collective influence as well, offering tenants collectively a say in the management of their estate and on general policy issues. Influence is encouraged through the provision of resources and other forms of support to tenants' groups.

Therefore, all other things being equal, one would have thought that tenants in citizenship authorities would be in a more powerful position than those in a consumerist authority. However, the evidence of this chapter, although far from conclusive, points to the fact that this is not necessarily so. First, all other things are not equal and many citizenship authorities are large, poor, inner city councils which are struggling to provide a satisfactory housing service. As some examples in the chapter showed, even if they agreed to give tenants what they wanted, they could not always deliver because of factors beyond their control. Second, a collective voice does not mean that tenants necessarily get what they want. A number of examples have shown that landlords can play the game in such a way that they will 'win'.

If those two factors, external events and playing the game, are taken into account then it is by no means certain that citizenship authorities can provide a service which tenants want. This is crucial because tenants have an instrumental view of participation and judge it mainly by what is achieved rather than the process. Thus in our research the consumerist authority, Sainsborough, was judged to listen to its tenants more than the citizenship ones in line with the perception that it produced a better service.

Clearly tenants can be helped in playing the game by having more available resources and the skills to organise and deploy them, and the citizenship authorities described here took many of these actions. Also, the formal structures of participation and the processes involved – the rules of the game – are clearly important in influencing the position of tenants. However, examples show how these local factors are not necessarily enough to ensure that tenants can receive the kind of housing management they would like or expect because of the influence of wider factors such as the political culture of the time and more specific factors such as the position of council housing as a whole.

8

CONCLUSION
Tenant–landlord relationships and the future of council housing

This book has argued that while the relationship between council tenants and their landlords has changed over recent years, in favour of greater tenant participation, the significance of this should not be exaggerated. Tenant activists rather than tenants generally, in some but not all areas, have been involved; the forms and purposes of participation have varied; some housing issues have been addressed more than others and, while tenant participation appeared to be growing, council housing itself was increasingly stigmatised and seen as a politically weak sector. The aim of this final chapter, therefore, is to consider the implications of the recent experience of tenant participation as outlined in the previous chapters. Key questions to be addressed are to what extent and in what way has the nature of the landlord–tenant relationship changed? What does this mean for the nature and future of council housing? This chapter will also focus on the factors which have stimulated and sustained the developments described in the book. Why has tenant participation become more popular among housing managers and councillors and will its growth continue? Finally, the implications of changes in council housing for the nature of local government in Britain are drawn out. It was mentioned in Chapter 1 that housing has traditionally been an important function of district and metropolitan authorities in Britain. Also, the point was made that developments in housing mirror changes in other local government services. Therefore, a key question is whether the changing role of council housing identified in this book gives an indication of what local government in general will look like in the future.

In Chapter 2, three ideal types of landlord–tenant relationship were outlined, and throughout the subsequent chapters the six local authorities in our study were described and compared to the ideal types. This chapter starts with an analysis of each of these ideal types in turn, focusing on the processes, objectives and impact associated with them. This is followed by an analysis of why tenant participation has grown over the past twenty years and an assessment of its future. Finally, the implications for the future shape of local government are briefly examined.

THE TRADITIONAL APPROACH

Of the three types, the traditional approach shows least change from well-established modes of local government functioning, based on concepts of professionalism and representative democracy. This is not to say that the extent of change from previous practice has been minimal. In Chapter 3 it was clear that very few housing managers, even in the traditional authorities, were prepared to say that tenants should have no say in the running of their housing. Similarly, few councillors were prepared to say that tenants should have no say in decisions which affected them. However, in traditional authorities both the extent and scope of collective tenant participation was limited to specific, usually estate-based issues. Little effort was made to support tenants' associations in general, although some support could be offered in particular problematic localities especially where there was some pressure from tenants. Also, the position of individual tenants as consumers was not well-developed. For example, none of the traditional authorities in our study carried out extensive market research or dealt effectively with tenant complaints. Therefore, in general the authority was reactive rather than proactive in its approach to dealing with its tenants.

The lack of any more developed form of tenant participation was explained by housing managers and councillors in three ways. First, it was said that there was no demand from tenants for further participation. Tenants in general were said to be not interested or only interested if there was a specific issue which affected them directly, such as modernisation of their own house and area. Second, some housing managers and councillors viewed as inappropriate the involvement of tenants' associations in questions of general policy. Their legitimate role was said to be helping to deal with certain practical problems on their own estates. Third, some councillors and housing managers said they were concerned about the legitimacy and representativeness of tenants' representatives and used this to justify not dealing with particular tenants' groups.

The first of these arguments, that there is no demand for participation from tenants, is clearly false on the basis of our research findings. In the traditional authorities, a large majority of tenants said that they wanted more say in decisions over housing management and there was considerable support for tenants' associations in principle, irrespective of whether they did or did not exist in practice. In two of the traditional areas the group discussions with tenants resulted in tenants present expressing interest in forming an association without any specific prompting and indeed to the embarrassment of the researchers. What the tenants lacked was the knowledge that other tenants thought the same way and information about how to set about forming an association.

The second argument, that tenants' associations only had legitimacy when dealing with practical, estate-based issues can be construed as housing managers trying to defend their professional status and councillors seeking to retain their traditional power. However, these views had been amended to some extent because of the acceptance by housing managers and councillors that tenants had knowledge of local situations which they did not have. Participation was viewed as a way of gaining information on local issues. Therefore, it did not challenge directly the housing managers' or councillors' claim to professional knowledge and political legitimacy. However, an involvement in general policy could be perceived as a direct challenge.

Some housing managers saw tenants' associations as a means of easing their own workload and of dealing with situations which they found problematic. Thus tenants' groups were seen as providing a useful function in channelling tenants' complaints and acting in other ways as a conduit of information. Also, associations were expected by some housing managers to take a leading role in estate management over problems such as vandalism or dumping of rubbish in unauthorised places. No such advantages were claimed for tenant involvement in general policy issues.

Thirdly, the view which councillors and housing managers sometimes held that tenants' associations lacked legitimacy was not shared by tenants themselves. Very few tenants thought that tenants' associations were unrepresentative of the views of tenants, although tenant activists did complain of the difficulties of getting more people actively involved. However, tenants questioned the legitimacy of councillors, with a large proportion disagreeing that it was up to councillors on their own to make decisions on housing management issues. Further, the legitimacy of councillors who did not live in council housing or did not have any experience of it was questioned by tenant activists.

However, tenants in the traditional authorities were more likely than those elsewhere to show a trust in traditional forms of democratic accountability and a belief that the council was looking after their interests. Therefore, it can be argued that the perceived legitimacy of councillors was greater in these areas than in the consumerist or citizenship authorities.

Consumerist mechanisms were also resisted on similar grounds as they were held to be unnecessary due to the perceived adequacy of traditional methods of political representation and of professionalism. Nevertheless, it was increasingly accepted that professionals needed information on what consumers needed or wanted in order to exercise professional judgement and that representative democracy was a very blunt instrument for dealing with individual consumer complaints.

Tenants of traditional authorities find themselves in a difficult power relationship with housing managers and councillors. Most tenants have to take up any issue on an individual basis and, as shown in the previous chapter, this gives them few resources to be able to influence the council. The tenants' associations which exist in areas where participation is tolerated or encouraged by traditional authorities do have some resources at their disposal. For example, if participation is around a modernisation programme, the authority needs the support of residents in order for the work to be carried out efficiently. Also, the local nature of the issue, directly affecting tenants, should make it relatively easy to organise tenants by enlisting their support. However, the lack of long term general support for tenants' associations is likely to mean that organisational resources and skills may not be well developed. It is fairly common for the tenants' associations to fold once the initial issue has been resolved, because of a lack of long term support, or sometimes the hostility which is directed towards tenants' representatives and the lack of a practical issue to progress which is deemed legitimate by the council. This can lead to an inability to meet tenants' objectives which can result in the disillusionment of tenant activists and questioning of their legitimacy by ordinary tenants.

Perhaps above all, tenants in a traditional authority have great difficulty in challenging the landlord's agenda and in getting their own issues discussed. The rules of the game are usually set by the landlords and in none of our traditional case study authorities was a tenants' group able to challenge this position. Therefore, in the traditional authorities participation is generally on the terms set by the landlord.

In these circumstances it is easy to be cynical about the motives of housing managers and councillors. It appears that participation is only tolerated if it meets their interests as they define them. Housing managers are looking to get tenants to make their lives easier; and councillors are anxious to be kept informed of any disquiet which could develop into a major issue leading to their political downfall at the hands of the voters. Nevertheless, it has to be emphasised that not all housing managers and councillors held these views and some were looking for an alternative approach which adopted more of the tenants' agenda. However, they were in a minority.

Both housing managers and councillors expressed a desire to keep tenants happy. At the time of our research the threat to council housing from Tenant's Choice seemed to be real; and the adoption of participation within the traditional approach was a reaction to this. However, this raised the question of how entrenched this relatively new element of the traditional approach is. If the perceived threat to council housing was lifted, it is possible that tenant participation would be jettisoned if the other advantages to the landlord were not considered important enough. However, in our research, these advantages were perceived as important by housing managers, councillors and tenants. Therefore, it is likely that some form of limited participation may remain, although, as we shall discuss later, it could be under some threat from the rise of consumerism and the adoption of consumerist mechanisms such as compulsory competitive tendering.

THE CONSUMERIST APPROACH

Only one authority in our sample was categorised as consumerist although elements of the approach were evident in both traditional and citizenship authorities. The particular circumstances of our consumerist authority are interesting in that it was a hung council. Immediately before the beginning of our research it had been Conservative-controlled although Labour was in power before that. In the hung situation there was no formal power sharing arrangement and, therefore, parties decided their voting intentions on each issue as it arose. As a consequence there was much uncertainty over the direction of policy and political control was relatively weak. The consumerist approach was partly a managerial initiative taken by officers. It represented a lowest common denominator between the three political parties trying for influence. The Conservatives supported consumerism, but were generally opposed to any collective influence for tenants as in the citizenship approach. The other two parties supported the citizenship approach and sometimes combined to implement aspects of it. However, they did not see the consumerist approach as being in contradiction to citizenship and so were happy to adopt consumerism as a stepping stone to a more collective approach.

Clearly the adoption of consumerism has emerged in Sainsborough in particular circumstances, but it is clear that it can emerge in other situations as well. For example, there are a number of authorities which seem to have adopted a consumerist approach following strong political direction from Conservative administrations. Some Labour-controlled councils have adopted consumerism with vigour although Labour is usually associated with a citizenship approach.

The consumerist view was usually associated with the rhetoric of the private sector. Tenants were seen as customers and considerable emphasis was placed on the fact that tenants pay for the housing service and, therefore, can be said to be the housing managers' employers. The use of this terminology was not confined to the consumerist authority as it was widespread among housing managers in all of the authorities. Even in the traditional authorities consumerist rhetoric was common, although it was not generally put into practice in any systematic way. However, in Sainsborough the ethos was pervasive and underlay much that the housing authority did. For example, there was considerable emphasis on discovering the views of tenants on the service provided, through survey research and the use of feedback on specific services through the use of prepaid postcards and so on. The authority made every effort to project a favourable businesslike image to tenants through written material, the behaviour of staff and the layout of offices. Above all the emphasis was on the delivery of a good and effective service.

Most interaction between the tenants and the landlord was at an individual level but the authority did not actively discourage the formation of tenants' groups. However, neither was the formation of groups actively encouraged in any systematic way, although it must be acknowledged that the previous history of tenant partici-pation in the area made it difficult for the council to build bridges with tenants' groups and convince them that it was acting in good faith. Nevertheless, the approach adopted by the consumerist authority stopped well short of the formal and usually complex structures of tenant participation adopted by citizenship authorities. Collective participation by tenants was primarily on an informal basis and was largely restricted to matters of service delivery rather than of policy. In this way the professionals retained a major influence over the nature of the service delivered while having the information to meet tenant needs.

This raises a fundamental question of the consumerist approach which is the power relationship between landlord and tenant. It can be argued that tenants are in a very powerful position in consumerist authorities because they have considerable influence over the service provided to them. Indeed in our study the consumerist authority provided the service which was rated most highly by tenants. It can be argued that the instrumental approach of most tenants means that it is the production of a good service which counts rather than the way it is achieved. Therefore, the lack of citizenship type participation structures is unimportant.

However, this argument cannot be sustained on the basis of the evidence of this study of only one consumerist authority. Its relatively good performance may be due to factors other than its consumerist orientation. For example, the authority did not have to deal to the same extent with the acute housing, economic and social problems which confronted the two citizenship authorities. Further, its stock of council housing was much smaller than those of the citizenship authorities and other research evidence shows that relatively small landlords perform more effectively than the very large ones (Clapham, 1992). It must also not be overlooked that the citizenship authorities pursued effective service delivery through consumerist mechanisms as well as employing a more collective participative approach. Therefore, for all these reasons, the link between consumerism and effective service performance has not been made effectively.

When looking at the power relationship between the landlord and individual tenants, which is the crux of the consumerist approach, it is hard to escape the conclusion that the position of the tenant as consumer is a very limited one. This is not to question the value of good information, market research and mechanisms of complaint and redress which clearly mean that the tenant is treated more like a consumer than a supplicant. Chapter 5 showed that paternalism has characterised the attitudes of housing managers and councillors to tenants in the past; and it appeared that this had been only slightly modified in the traditional authorities studied here. A key aspect of the position of consumers is the ability to exercise choice. In purchasing private commodities, consumers usually have the opportunity to choose between different products or brands, and there is a regulatory system designed to ensure that the consumer has this choice. Of course one can query the degree of choice which consumers sometimes have in practice in choosing between a few products which are almost the same except for their brand image. Nevertheless, consumers usually have the sanction of withdrawing their custom if they are dissatisfied.

In essence council tenants do not have this sanction, although at the time of our research it seemed to housing managers at least, that Tenant's Choice would provide this. However, this mechanism has not lived up to the initial expectations, and tenants in general do not seem to have been impressed by the opportunity to choose a private landlord or a housing association. Also, the mechanism itself is a flawed way of offering choice, because it involves a once and for all transfer, and changes ownership, as well as the management service (Clapham, 1990). CCT offers a mechanism for changing the housing management agency, but the choice is made by the council rather than by the tenants, although it is possible for tenants to be involved in the decision (Goodlad and Rosengard, 1994).

Above all, the limitation of consumerism in housing relates to the difference between housing and other commodities. As a number of tenants quoted in Chapter 5 pointed out, renting a house is not like renting a television, because you cannot just take it back if you are not satisfied. Housing is a commodity which is fundamental to people's lives and which influences access to other public and private services such as health and education, and to employment. The tenant as consumer has few resources which can be deployed to bring influence to bear on the landlord. Gyford (1992) states:

> In the final analysis there is perhaps something rather limited in the concept of the local consumer. Consumption is an act of receipt rather than of creation and no matter how much their complaints are listened to or their interests are taken into account, consumers are ultimately a means to the wider ends of their providers, be they in the public or private sectors, be they political or financial.
>
> (Gyford, 1992, pp.168–9)

Two points in this quotation are worthy of consideration. The emphasis on the passive role of receipt rather than creation is contentious because it clashes with the expressed instrumentalism of most tenants, who do not seem to mind being passive if they receive a good service. Also, it can be argued that the views of consumers, if gleaned through market research, can lead to the creation of new services in the future. In this sense only, the role is not just a passive one. However, the second point concerning the manipulative relationship involved in consumerism is easier to sustain, because of the inability of tenants to set the agenda or to deploy resources to effectively challenge the landlord.

Gyford continues his argument by stressing the advantages to councils and other service providers of the consumerist model.

> The public confined to individual roles may be easier to handle than the public as actual or potential groups making organised demands and intruding as a political force into the policy making process ... The public acting as private individuals rather than as organised groups may be less threatening to established centres of power even if they do require the expenditure of extra time and effort to respond to their demands.
>
> (Gyford, 1992, p.184)

The managerial basis of consumerism in our local authority was mentioned earlier. It did not arise from tenant demand, but from a political vacuum where managers were freed from detailed day-to-day control, and indeed received little guidance on strategic direction. Therefore one interpretation of the adoption of consumerism in these particular circumstances is that it is a strategy to preserve managers' positions in the face of perceived external threats from Tenant's Choice and other central government mechanisms designed to reduce the role of council housing. It can improve the standards of service and keep tenants happy while not surrendering managerial autonomy or control over the service. Consumerism can also be seen as a way of reinforcing the housing managers' position by depoliticising' housing management and treating it as essentially a technical and managerial rather than political task. Later moves to transfer the stock to a new housing association run by existing housing management staff reinforce this interpretation.

However, the political situation in this authority and elsewhere should not be ignored. The consumerist approach fits with Labour and Liberal Democrat ideas to improve services even though it stops short of the citizenship ideal. Also, some Conservative councils have pursued consumerism as it is seen as being congruent with the philosophy of privatisation. Therefore the development of consumerism is a fusion of managerial and political pressures.

The final question concerning consumerism is whether it is a lasting phenomenon or merely a short stage in a transition to something else. Although it is difficult to answer the question in any definitive way, there are doubts about its long term sustainability. For Conservatives and managers supporting consumerism, the logical progression is to take the route suggested in Sainsborough of removing the housing service from local authority control. If housing is a commodity which can be marketed like any other, then there is no logical reason why it should be provided or managed by local authorities rather than voluntary or private sector organisations. Therefore, consumerism as a part of the management of council housing may be only a transitional phenomenon. We shall return to this issue later in the chapter.

THE CITIZENSHIP APPROACH

The third ideal type, citizenship, was pursued in two of our case study authorities, both large urban councils controlled by the Labour Party. The approach was justified on a number of grounds by housing managers and councillors. First, it was argued that tenants had a right to a good standard of service and that this could be achieved by the adoption of consumerist ideas as well as more collective involvement. The latter was important in adding to the feedback from tenants which landlords received through individualist methods. It was argued that a process of dialogue was necessary if the service was to meet the needs of the tenants. Second, tenants were said to have a right to be involved in the running of the houses they lived in, and the best way of achieving this was through the collective organisation of tenants and their involvement in formal participative machinery. The tradition of tenant organisation – which these authorities fostered – had a long history and was seen to be related to the Labour movement which controlled these authorities. Third, there was a perception that the authorities and the tenants' movement had a mutual interest in standing together in opposition to government policy.

Therefore, the arguments for the approach are a mixture of a view held by councillors and housing managers of the appropriate political rights of tenants and a perception of an appropriate way of providing a good service to tenants as well as a way of mobilising against a common enemy. In Compton and Strathmuir tenants' representatives supported the citizenship approach and had clearly built up a good relationship with their landlord, although this did not mean that they were uncritical. Tenants in the household survey supported the idea of collective involvement by tenants but were no happier than in other authorities about the degree to which the landlord listened to their views or acted in their interests. The lack of favourable impact of the citizenship approach on tenants as a whole raises two important issues.

The first is that the collective approach in the two authorities involved only a very small proportion of tenants in the participation process. The tenants' representatives gained in skills, knowledge and confidence and became aware of the factors operating on the local authority which constrained the landlord's ability to meet tenants' demands and needs. However, these impacts were not felt by tenants as a whole.

The second, related point is that the standard of service which was provided in the two citizenship authorities was perceived by tenants as being less satisfactory than in any of the traditional or consumerist authorities. In other words, in the eyes of tenants the citizenship approach did not deliver the standard of service they wanted. However, this must not necessarily be taken as an indictment of the citizenship approach as the factors which created the poor performance may be outside the control of the local authority. The citizenship approach has been associated with the Labour Party which generally controls large urban authorities with pressing social problems. Compton and Strathmuir were typical of such authorities and contained substantial concentrations of deprivation and poverty in their areas. A number of research studies have shown that the problems of size of stock and lack of finance have resulted in these authorities providing a poor standard of service to tenants (for a review of the evidence, see Clapham, 1992a). Therefore, citizenship has been adopted by authorities with a poor standard of performance rather than the poor performance stemming from the citizenship approach.

Adoption of the citizenship approach clearly, by itself, does not turn a bad service into a good one. Also, it does not change the perceptions of tenants as a whole of the landlord or the quality of service provided. Therefore, it is possible that the citizenship approach itself may be a transitory phenomenon as it is seen not to achieve the objectives of councillors, housing managers and tenants. However, there was little evidence of this perception in Compton and Strathmuir where there remained a strong commitment to the ideals of the citizenship approach.

The citizenship approach was the one which represented the most direct challenge to the traditional status of both councillors and housing managers. Some were prepared to accept this challenge for political and practical reasons. However, others were not, and this could be a limiting factor in the spread of the citizenship model.

The research reported here did dispel one doubt about the citizenship approach, which is that tenants were not interested in becoming involved. Clearly not many tenants did want to get directly involved in tenant participation themselves, but crucially they supported the concept of collective involvement and expressed support for and confidence in tenants' representatives who often spend a lot of time in this role.

The key question of the citizenship approach is what difference it makes to the power relationship between tenants and their landlord. The lack of impact of the approach on the poor standard of service provided would seem to render ineffective the time put in by tenant activists whose role often exposed them to criticism from tenants. In Chapter 6 it was noted that tenant activists often defended housing managers who they felt were being unjustly attacked by tenants, and developed an awareness of the constraints which influenced service delivery. Therefore, is a citizenship approach by councillors and housing managers able to achieve any substantive improvements in tenants' housing conditions?

Tenant activists themselves did not feel that they were being manipulated and incorporated and pointed to victories and concessions which they had achieved in the face of opposition from their landlord. Examples were given in the previous chapter of how organised tenants had managed to set the participation agenda and attain their objectives. Undoubtedly the citizenship approach gives the opportunity to organise their resources and to exert pressure on other players in the game where necessary. Nevertheless, the other players have substantial resources of their own and external constraints on the game may limit what can be achieved. This makes the outcome of tenant participation less predictable than in the other types of authority. More positively, this means tenants have a better chance of influencing the policy process than in the other authorities.

CONCLUSION

Citizenship therefore reflects as well as creates conditions in which tenants may achieve greater relative power in their relationship with their landlord. But given the fluidity in internal relationships, and the unpredictability of external constraints, it is not possible to predict the outcomes in particular cases.

Citizenship may not offer tenants much in the way of improvements in the general service they receive, but neither does it offer great disadvantages apart from the time spent on participation activities by tenant activists. They perceive their efforts as being rewarded by myriad small improvements on their estates and in the situation of individual tenants being improved as well as in the occasional major victory by changing policy.

The limitation of the citizenship approach, with its emphasis on formal participation structures, is that the impact on the ground is limited both in the service provided and in the involvement and perception of tenants. However, the citizenship approach can also involve the creation of community-based organisations such as co-operatives which are run by tenants. Other research has shown that the creation of co-operatives is effective in improving the standard of service provided to tenants and in providing a mechanism for widespread tenant involvement (Clapham and Kintrea, 1992). Both of the citizenship authorities in our study supported to varying degrees the creation of such organisations. However, relatively few had been created and so their overall impact was small.

In conclusion, all three approaches involve a change from previous ways of managing council housing, but the extent of change varies considerably. The traditional model is largely a minimal response to a perceived outside threat, which does little to improve the resources available to tenants to change the relationship with their landlord. The consumerist approach is the one being advocated strongly by central government. Its emphasis is on the improvement of service delivery and on the participation of individual tenants largely through the exercise of choice and the interchange of information. However, it is clear that this approach gives only limited resources to tenants with which to influence the landlord. It does not represent any great change in the power relationship between landlord and tenant. The landlord as producer makes all major decisions and is constrained only marginally by the need to take into account tenant views, largely because the mechanisms of 'exit' such as Tenant's Choice are difficult to implement in practice and 'voice' mechanisms are limited.

The citizenship approach offers tenants collectively a voice in housing management but, those who participate feel, as argued earlier, that their achievements may sometimes be very limited. Further, the approach does not appear to make a major impact on tenants outside the formal participation machinery. However, many of the reasons for this lie outside the control of the landlord or its tenants.

WHY HAS TENANT PARTICIPATION INCREASED?

In the light of the varied nature of tenant participation outlined above, this chapter continues by examining the factors which lie behind the increase in arrangements for tenant participation. In Chapter 7 an analysis was offered of the changing context of tenant participation in the early 1970s and the early 1990s. It was emphasised that much tenant participation in the 1970s was stimulated by pressure from tenants who campaigned actively, particularly over the issue of increasing rents and the imposition of 'fair rents'. The importance of the issue and its direct relevance to tenants as a whole meant that tenants could be mobilised relatively easily. At the same time, local authorities and successive central governments were in weak political positions and felt themselves vulnerable to sustained tenant pressure.

However, in the mid-1990s the situation has changed. There is no one issue which unites tenants and the rents issue has been written out of the rules of the game and defused through the introduction of housing benefit. In addition, the political importance of council tenants has declined as the sector has reduced in size and marginal social groups have increasingly been concentrated in it. Therefore, the context within which tenant participation takes place has made it more difficult for tenants to achieve their collective objectives and also for tenants to recognise a common interest.

The transition from relative strength to relative weakness in the position of tenants between the 1970s and 1990s has coincided with the growth of arrangements for tenant participation associated with the citizenship and the traditional model. Therefore, this raises a number of questions. First, has the growth of tenant participation contributed to the relative weakness of the tenants' movement? Second, why have arrangements for tenant participation become more widespread when the tenants' movement is relatively weak?

It is very difficult to sustain the view that the spread of arrangements for tenant participation has enfeebled tenants. The argument is based on the two premises that tenants are incorporated into a system of participation which offers them little influence in return and further, that tenants have given up influence to engage in participation.

On the first premise it has been shown in Chapter 7 that tenants can achieve some influence through tenant participation arrangements, both in winning 'games' over particular issues and in changing the rules of the game. Clearly this does not mean that tenants 'win' on every occasion, or even that every group of tenants wins even once. Nevertheless participation structures and processes can be of assistance to tenants in achieving their objectives. The objectives of some councillors and housing managers have been to use tenant participation to legitimise existing policies or to encourage tenants to make their work easier by taking on certain tasks, or even as a way of attempting to change the attitudes of tenant activists. But they have not always achieved these objectives as they are confronted by tenants with other objectives and resources to organise in order to achieve them.

The second premise that tenants give up something to engage in tenant participation is questionable. The assumption is that tenants achieve their objectives through militant action. On the rents issue in the early 1970s this view may be justified but there is little evidence in the mid-1990s that tenants are interested in or prepared to take militant action as was shown in Chapter 6.

But crucially tenants do not necessarily give up this option by being involved in participation. Indeed, it was accepted by many councillors and housing managers in traditional and citizenship authorities that tenants were 'within their rights' to use whatever measures they felt were appropriate to achieve their objectives once the participation channels were exhausted. Thus recourse to militant activity, although seen as a last resort, was considered to be part of the unwritten 'rules of the game' of participation and had been used in our case studies on a number of occasions with mixed results. There was no evidence that involvement in participation discouraged tenants from taking militant action where they felt it necessary. Indeed, the experience of participation seemed to give tenants the resources and confidence to take such action. It could be argued that the existence of tenant participation contributed to the clear reluctance of tenants to take any form of militant action, as shown by the results of the tenant survey. However, the reluctance was greater in the traditional authorities where little participation existed than it was in the citizenship authorities. Nevertheless, the issue of tenants' perceptions of their own influence and the most appropriate ways of pursuing it is an important one which deserves further analysis. Clearly the factors which influence tenant perceptions are many and their interaction complex. There is no evidence of a clear link to tenant participation arrangements.

Therefore there is little evidence that the growth of tenant participation has contributed to the weakening of the tenants' movement. On the contrary, the argument that participation has strengthened the tenants' movement by providing financial resources, experience and skills which can be organised to further tenants' objectives is a persuasive one, although one difficult to prove.

The second main question was why participation arrangements had increased, reflecting the growth of the citizenship model and the incorporation of tenant participation into the traditional model. The growth of arrangements for participation immediately followed the defeat of the tenants' movement over the rents issue in the 1970s and has continued at a time when the tenants' movement has been relatively weak. Although we have argued that tenant participation has not weakened the tenants' movement, this does not mean that the intention of some housing managers and councillors has not been to introduce tenant participation for this purpose. Some have clearly sought to use tenant participation to influence the tenants' agenda and to seek to offload some of their responsibilities onto tenants. However, these objectives did not seem to be widely shared in our case study local authorities. Also, the lack of success of tenant participation in achieving these aims compares with the obvious success which local authorities can have by failing to support tenant organisation. In the consumerist authority, and in one of the traditional authorities, refusal to aid tenants' groups or speak with them resulted in making the tenants' voice inaudible. So why turn to the more ineffectual means of tenant participation to weaken tenants?

The major reason for the rise of tenant participation since the 1970s has been the growing perception among housing managers, councillors and tenants that tenant participation was in their interests. A number of factors have been important in stimulating this recognition.

Perhaps the most important factor has been the residualisation of council housing (Clapham and English, 1987; Forrest and Murie,

1983). The impact of the growing concentration of socially marginal people in council housing and the restrictions on capital expenditure on the stock have combined to create severe housing management problems in many areas. The problems have been exacerbated from the 1980s onwards by the growing polarisation of society and the creation of a large group of people with little or no access to employment, a feature which has been reinforced by the long recessions of the early 1980s and early 1990s. The extent of poverty in 1990s Britain has been well documented (see for example, Oppenheim, 1990; Frayman, 1991; Cohen et al., 1992) and there has been increasing concern over the creation of an underclass cut off from the mainstream of society (Murray, 1990). Whether one accepts the underclass diagnosis or not, it is clear that many households living in many council housing areas experience a wide range of social problems including poverty, vandalism, crime, drug abuse and so on. These are exacerbated by a lack of social facilities and poor and deteriorating housing conditions.

Not all council housing areas are like this, but there is a tendency for tenant participation to be most well-developed in these areas. Both our citizenship authorities were large urban authorities where social problems were concentrated. In addition, traditional authorities have generally established participation structures in the problematic parts of their stock.

The social and housing problems have created severe housing management problems which can only be effectively tackled through an alliance of landlord and tenants. Action by the landlord is itself constrained by financial shortages and can be subverted or nullified by tenant opposition. Both parties have increasingly perceived a common interest in tackling housing management problems.

The possible option for landlords of doing nothing in such circumstances has been made risky by the political attack on council housing by central government, even if landlords were prepared to take the political risk and the moral responsibility for inaction. The perceived need to defend council housing whether for political reasons or for the maintenance of employment opportunities has meant, in many instances, that both housing managers and councillors have perceived their interests to lie in attempting to deal effectively with housing management problems. Tenants are of course largely motivated by the desire to improve the conditions they face, and the deterioration of council housing may have added impetus to the motivation.

The speed of the residualisation process has been determined by the implementation of political ideology. At the same time residualisation has politicised the management of council housing, in a way that was not evident before the 1970s. Therefore, the development of the citizenship and consumerist models has been given impetus by the application of political ideology to the management of council housing. Also, political ideologies are not fixed, but change over time. Thus, the citizenship model has arisen partly because of rethinking on the left over the relationship between the state and civil society and the adoption of citizenship as a general foundation of social democracy. Consumerism arose also to some extent because of the rise of 'New Right' ideology in the Conservative Party and the consequent application of public choice concepts and ideas to the public sector.

We have focused so far on the spread of tenant participation arrangements usually associated with the citizenship and the traditional models. However, with some exceptions, consumerism does not use the same arrangements and has been a relatively recent phenomenon occurring only since the early 1980s. At some levels consumerism may be seen as antagonistic to the other models and its growth may mean that the peak of the growth of collective participation arrangements has been reached. At other levels it is clear that the other models may be able to absorb the rise of consumerism. Citizenship ideas accept that consumerist rights and obligations are to be met and in practice many local authorities have espoused citizenship and consumerist ideas simultaneously. Also, the traditional model may be changed in practice to incorporate consumerist ideas and policy prescriptions, in particular those emanating from central government. Thus the incorporation and strengthening of the consumerist element of these models seems inevitable if central government policy continues in its present course.

However, some of the consumerist mechanisms which have been put forward such as CCT may pose a direct challenge to the collective participation arrangements of citizenship. At the very least the nature of the arrangements will have to change substantially if the citizenship ideals are to be met (Goodlad and Rosengard, 1994). It is possible that the new contractual arrangements and the system of incentives they create for the parties involved serves to reduce or even negate the incentive to be involved in tenant participation which we have noted above. The precise rules governing CCT may change, but it seems that tenants in most areas will have only a minimal role, if any, in choosing between different housing management contractors. The tenants' ability to influence housing management policy may be limited to the time when contracts are renewed and a tender brief is drawn up. Once the contract is awarded, tenant influence over housing management issues depends on the details of the contract specification. However, tenant participation is not an area which lends itself easily to unambiguous specification. How can you stipulate the processes of tenant participation (such as listening or dialogue) in a way which can be easily monitored? Attempts can be made to specify these processes and some performance indicators are available (see Cairncross, Clapham and Goodlad, 1989) but it is difficult to specify them in such a way that a potentially recalcitrant contractor can be held to account. It seems likely that there will be substantial pressure on contractors to keep costs to a minimum and the consumerist approach may prove desirable to them in these circumstances. The provision of a good service at least cost may take precedence over ideals concerning the collective rights of tenants. Relationships become more complicated with the client–contractor split inherent in CCT. Tenants' associations have to attempt to influence both the client and the contractor. Although the citizenship model is not necessarily inapplicable in this situation it is potentially more complicated and difficult for tenants to handle.

The consumerist model looks set to dominate the short term future of council housing management at the time of writing. However, the three models are the product of conflicting social and political forces, and radical change in a short period is possible. A change in the party in power in central government, for example could have a large impact on the relative balance between the models in the future. Also, the changing context within which social and political battles are fought may well mean that the nature of the models themselves will change. Therefore, it is extremely difficult to predict the future of tenant participation.

It has already been noted that consumerism could be transitory as a feature of council housing management. This is because the logic of the consumerist position is that council housing can be just as well owned or managed by private or voluntary agencies. If housing is a commodity like any other then there is no logical reason why it should be run by local councils. Clearly, the transfer of housing from local authorities is not a straightforward matter or one which can be achieved quickly. Therefore, council housing will exist for some time to come. Nevertheless, the combined impact of Right to Buy sales, Tenant's Choice, voluntary transfer, CCT, housing companies and support for housing associations and private landlords could, over a period of time threaten the existence of council housing. However, in the short term, consumerism in council housing seems set to increase in influence, permeating the traditional approach and being given added emphasis within citizenship.

The citizenship approach is clearly threatened by some developments, in particular CCT. But it is also vulnerable because of its lack of success in providing a good standard of service, even though this may largely be due to factors outside the control of landlords or tenants. A further limitation of the approach is its apparent inability to make an impact on tenants as a whole rather than on a small group of tenant activists. The combination of these two failures could make the approach difficult to sustain in the face of the strong push towards consumerism. Two developments in housing policy offer an alternative way forward for council housing which would accommodate citizenship more easily as well as consumer ideas. Perhaps the best defensive measure would be to change the emphasis within citizenship by stressing collective control through tenant management organisations. This strategy counters the threat of CCT and creates strong citizenship structures at a very local level which can engage more tenants. There is not space here to discuss in depth the reason for such a strategy or its likelihood of success (for further details, see Clapham, 1989). However, it has the danger of confining tenant influence to estate-based issues unless matched by authority-wide mechanisms of tenant involvement.

The second development is the idea of local housing companies, with strong representation on their boards from councils as well as tenants. They have strong support from both left and right in the political spectrum (Commission on Social Justice, 1994; Raynsford, 1992; Department of the Environment and Welsh Office, 1995) as well as from professionals and academics (for example, Wilcox et al., 1993). The debate about local housing companies has been dominated by issues of finance, but if they are to provide a truly alternative form of housing which combines the most popular features of other forms of rented housing then the debate must move on to consider more fully how citizenship ideas can be incorporated in the model. They could provide an opportunity to develop authority-wide mechanisms of tenant involvement based on citizenship as well as consumerist ideas. If to this was added the emphasis on local or neighbourhood forms of participation which characterises the citizenship approaches, and the emphasis on individual, consumerist approaches, reviewed in this book, then the local housing company model would become a more comprehensive and attractive alternative to council housing.

The greater capital investment that local housing companies could engage in would also address tenant participation based on the citizenship model.

CHANGING LOCAL GOVERNMENT

The general picture in housing is complex and constantly changing but it nevertheless reflects the developments in other local government services which were outlined in Chapter 1. A number of key trends have been identified. First, there is the centralisation of policy. It is noticeable in tenant participation that central government has been increasingly involved in the 'rules of the game'. Examples were given in Chapter 7 of agreements between tenants' associations and landlords being impossible to implement because of financial constraints from central government. At the same time, central government has intervened more strongly since 1980 in the detail of housing management policy. Also it has used mechanisms such as the Estate Action programme in England to attach tight strings to local authorities to influence their spending priorities.

Second, centralisation has been associated with the rise of citizenship arrangements promoted by local authorities for involving citizens in the management of services. In council housing these arrangements were to some extent a reaction to central policy towards council housing and were attempts to form a common front against the 'enemy'. However, central government itself has pushed some mechanisms associated with citizenship. For example, the availability of Estate Action capital spending consent was dependent on, among other things, the establishment of structures such as an estate management board or other forms of local tenant involvement. Also, to some extent it can be argued that citizenship arrangements would have arisen anyway as they stem from the application of a particular political philosophy and were being introduced before current centralisation trends were fully established.

The third major trend has been the introduction of consumerism in public services, as seen in mechanisms such as Tenant's Choice. Consumerism has become enshrined in the culture of local government and is sometimes reinforced by performance assessment mechanisms as well as the imperative to compete.

The relationship between consumerist and citizenship trends is a very complex one. Central government could be seen as promoting both, and many citizenship authorities view consumerism as part of their approach despite objecting to some mechanisms such as CCT. Individual mechanisms are therefore difficult to ascribe to either a consumerist or citizenship approach. For example, central government has pushed tenant management co-operatives and estate management boards (as well as school boards, for example) as have citizenship authorities. The apparent paradox is solved through the realisation that individual mechanisms can have a wide range of objectives and can be used in pursuit of both approaches. Thus, one of the objectives for tenant management co-operatives is clearly the consumerist one of breaking the local authority monopoly and ensuring competition. Also, it can be argued that tenant management co-operatives embody the citizenship approach because of their collective and democratic nature. But they are essentially collectives of consumers and so they are not able by themselves to achieve the goals of citizenship which would require other additional participative mechanisms for citizens and non-consumers. As groups of consumers, tenant management co-operatives could be (and are) incorporated into a consumerist approach. Similar consumerist mechanisms within other local government services such as grant maintained schools have not attracted support from Labour's citizenship local authorities. But the Labour Party at national level is committed to developing such consumerist mechanisms into combined citizenship and consumerist models.

Likewise, mechanisms such as CCT which at first glance may seem to be consumerist in their orientation are not necessarily so. For example, many housing co-operatives carry out their housing management functions by awarding contracts on a competitive basis to outside agencies. Partly for this reason they have not always been favoured by Labour local authorities.

These individual mechanisms should not be judged in isolation, but rather in the context in which they are situated. With this in mind it is clear that CCT as it has been implemented in British local government reinforces a consumerist approach and makes the achievement of a citizenship approach more problematic, largely because of the detailed procedures which stipulate who makes the decision over which tender to accept and which criteria are used.

Thus, citizens are likely to have little say in decisions and the criteria are weighted towards consumerist considerations of efficient and low cost services rather than citizenship ideas of rights and obligations. The conclusions of this book relating to changes in housing cannot be applied directly to other local government services. However, the analytical framework used here is directly relevant to the study of changes in other services and we would contend that the general models of traditionalism, consumerism and citizenship are applicable to local government in general. The nature of local government has changed as the models have developed and the emphasis between them has changed.

This is seen in other local government services such as education. Central government has been more involved in schools through the national curriculum and the direct funding of opted out schools. As well as the management of the latter, parents have been involved in the running of local authority schools through school boards. The introduction, by the government, of league tables of school perfor-mance has been aimed at giving information to parents to exercise choice between schools. Le Grand and Bartlett (1993) have termed this new structure of provision a 'quasi-market' and have described its impact in a number of local government and other public services.

The emphasis in the government's approach is on competition both between public agencies and with the private sector. A primary aim is to ensure efficient services. For example, Ridley, a former Secretary of State responsible for local government argues that 'what the general public wants of local government is that it should provide good services as efficiently as possible' (Ridley, 1988, p.11). Similar objectives are expressed by the Right for mechanisms such as CCT.

The emphasis on competition is aligned with a view of account-ability which is seen in terms of consumer choice rather than political representation. As Stoker argues this means that:

> Collective control through an elected local authority is under challenge from a Conservative programme which hands over control of key services to sectional interests and more gener-ally stresses the role of individual consumer participation in a market place of the public sector.
>
> (Stoker, 1989, p.165)

The development of quasi-markets, coupled with consumerism, therefore carries the potential to eliminate formerly monopolistic local government services. This direction in policy was so well established by the early 1990s that the elimination of council housing was widely predicted to follow the general confirmation in the direction of policy represented by the 1995 White Paper (Department of the Environment and Welsh Office). Of the three models for tenant–landlord relations discussed in this book, consumerism was widely perceived to have become dominant.

The difference between consumerism and citizenship is essentially the difference between local administration and local government. Stewart argues that the fragmentation which is the necessary consequence of competition could make it difficult to pursue a proactive government role, 'Local government will be lost between the fragments' (Stewart, 1989, p.179). Therefore, if consumerism continues to take over the town halls and the housing offices, the citizenship version of local government will not be developed.

The battles between the different models in housing reflect battles between different views of what local government in general should look like. Should it be primarily the provider of efficient services to individual consumers, or should it be primarily the focus of a network of citizenship rights and obligations? As in housing, it is difficult to predict the direction of change in local government for anything longer than the immediate short term, when it is clear that consumerism will be increasingly influential.

Yet many signs of a continuing commitment to citizenship ideas are apparent in local government and elsewhere. Whether this means a future for council housing under a New Labour government is not clear. It may be impossible to develop a satisfactory model for council housing and tenant–landlord relations while keeping the housing stock within the public sector as defined for public spending purposes. Whether the future lies in council housing or in some combination of local housing companies, housing associations and housing co-operatives, an opportunity exists to develop a model for public rented housing and tenant–landlord relations which builds on the collective strengths of council housing and citizenship ideas, and combines these with a commitment to consumerist ideas about quality of provision and efficiency. The tensions between citizens and consumers which this model implies would be reconciled in the playing out of 'the game' of tenant participation, involving tenants, housing managers, councillors and others engaged in the task of creating and reproducing the power relationships of the public rented sector.

REFERENCES

Almond, G. and Verba, S. (1965) *The Civic Culture*, Boston: Little, Brown

Andrews, L. (1979) *Tenants and Town Hall*, London: HMSO

Audit Commission (1986) *Managing the Crisis in Council Housing*, London: HMSO

Bachrach, P. and Baratz, M. (1963) 'The two faces of power', *American Political Science Review* 57, pp. 641–55

Barron, J., Crawley, G. and Wood, T. (1991) *Councillors in Crisis*, London: Macmillan

Bines, W., Kemp, P., Pleace, N. and Radley, C. (1993) *Managing Social Housing*, London: HMSO

Birks, D. and Southan, J. (1992) 'An evaluation of the rationale of tenant satisfaction surveys', *Housing Studies* 7, pp. 299–308

Bloch, A. (1992) *The Turnover of Local Councillors*, York: Joseph Rowntree Foundation

Brion, M. and Tinker, A. (1980) *Women in Housing*, London: Housing Centre Trust

Buchanan, J. and Tullock, G. (1962) *The Calculus of Consent*, Ann Arbor: University of Michigan Press

Cairncross, L., Clapham, D. and Goodlad, R. (1989) *Tenant Participation in Housing Management*, Coventry/Salford: Institute of Housing/Tenant Participation Advisory Service

Cairncross, L., Clapham, D. and Goodlad, R. (1990) 'The pattern of tenant participation in council housing management', *Discussion Paper No. 31*, Glasgow: Centre for Housing Research

Cairncross, L., Clapham, D. and Goodlad, R. (1993) 'The social bases of tenant organisation', *Housing Studies* 8, pp.179–94

Central Housing Advisory Committee (1939) *Management of Municipal Housing Estates*, London: HMSO

Central Housing Advisory Committee (1969) *Council Housing, Purposes and Priorities*, London: HMSO

City University (1977) *Housing Training: Education and Training for Housing Work Project*, London: City University

Clapham, D. (1987) 'Trends in housing management', in Clapham, D. and English, J. (eds) *Public Housing: Current Trends and Future Developments*, London: HMSO

REFERENCES

Clapham, D. (1989) *Goodbye Council Housing?*, London: Unwin Hyman
Clapham, D. (1990) 'Housing', in Deakin, N. and Wright, A. (eds) *Consuming Public Services*, London: Routledge
Clapham, D. (1992a) 'The effectiveness of housing management', *Social Policy and Administration* 26.3, pp.209–25
Clapham, D. (1992b) 'A woman of her time', in Grant, C. (ed.) *Built to Last: Reflections on British Housing Policy*, London: Roof
Clapham, D. and English, J. (eds) (1987) *Public Housing: Current Trends and Future Developments*, London: Croom Helm
Clapham, D., Kemp, P. and Smith, S. (1990) *Housing and Social Policy*, Basingstoke: Macmillan
Clapham, D. and Kintrea, K. (1986) 'Rationing, choice and constraint: the allocation of public housing in Glasgow', *Journal of Social Policy* 15.1, pp.51–67
Clapham, D. and Kintrea, K. (1992) *Housing Co-operatives in Britain: Achievements and Prospects*, Harlow: Longman
Clapham, D., Kintrea, K., Malcolm, J., Parkey, H. and Scott, S. (1995) *A Baseline Study of Housing Management in Scotland*, Edinburgh: HMSO
Clarke, M. and Stewart, J. (1991) *Choices for Local Government for the 1990s and Beyond*, Harlow: Longman
Clegg, S. (1989) *Frameworks of Power*, London: Sage
Clinton, A., Murie, A., Plaice, D., Tolan, F. and Williams, P. (1989) *The Relative Effectiveness of Different Forms of Housing Management in Wales*, Cardiff: Welsh Office
Cohen, R., Coxall, J., Craig, G. and Sadiq-Sangster, A. (1992) *Hardship Britain: Being Poor in the 1990s*, London: Child Poverty Action Group
Cole, I. and Furbey, R. (1994) *The Eclipse of Council Housing*, London: Routledge
Commission on Social Justice (1994) *Social Justice: Strategies for National Renewal*, London: Vintage
Committee on Local Authority and Allied Personal Social Services (1968) *The Seebohm Report*, Cmnd 3703, London: HMSO
Craddock, J. (1975) *Tenant Participation in Housing Management*, London: ALHE
Dahl, R. (1961) *Who Governs: Democracy and Power in an American City*, New Haven and London: Yale University Press
Damer, S. (1974) 'Wine Alley – the sociology of a dreadful enclosure', *Sociology Review* 22.2, pp.221–48
Dearlove, J. (1973) *The Politics of Policy in Local Government*, London: Cambridge University Press
Department of the Environment (1987) *Housing: The Government's Proposals*, Cm. 214, London: HMSO
Department of the Environment (1990) *Tenants in the Lead*, London: HMSO
Department of the Environment (1992) *Local Government Review: The Functions of Local Authorities in England*, London: HMSO
Department of the Environment (1994) *Access to Local Authority Housing: A Consultation Paper*, London: DoE
Department of the Environment and Welsh Office (1995) *Our Future Homes: Opportunity, Choice, Responsibility*, Cm. 2901, London: HMSO

REFERENCES

Dunleavy, P. (1980) *Urban Political Analysis*, London: Macmillan

Durant, R. (1939) *Watling: A Survey of Social Life on a New Housing Estate*, London: P.S. King

Forrest, R. and Murie, A. (1983) 'Residualisation and council housing: aspects of changing social relations of housing and tenure', *Journal of Social Policy* 12, pp.453–68

Forrest, R. and Murie, A. (1988) *Selling the Welfare State: The Privatisation of Public Housing*, London: Routledge

Forrest, R. and Murie, A. (1995) (eds) *Housing and Family Wealth*, London: Routledge

Frayman, H. (1991) *Breadline Britain 1990s*, London: Domino Films and London Weekend Television

Gallagher, P. (1982) 'Ideology and Housing Management', in English, J. (ed.) *The Future of Council Housing*, London: Croom Helm

Goodlad, R. and Rosengard, A. (1994) 'Tenant involvement in Compulsory Competitive Tendering', in *Housing Management Compulsory Competitve Tendering*, London: Association of District Councils and Chartered Institute of Housing

Gyford, J. (1984) *Local Politics in Britain* (Second Edition), London: Croom Helm

Gyford, J. (1992) *Councils, Citizens and Consumers*, Basingstoke: Macmillan

Gyford, J., Leach, S. and Game, C. (1989) *The Changing Politics of Local Government*, London: Unwin Hyman

Hague, D. (1990) 'The development and politics of tenant participation in British council housing', *Housing Studies* 5, 242–56

Hambleton, R. (1988) 'Consumerism, decentralisation and local democracy', *Public Administration* 66, pp.125–47

Hampton, W. (1970) *Democracy and Community*, London: Oxford University Press

Hampton, W. (1991) *Local Government and Urban Politics* (Second Edition), Harlow: Longman

Hills, J. (1995) *Inquiry into Income and Wealth*, Vol 2, York: Joseph Rowntree Foundation

Housing Services Advisory Group (1978) *Organising a Comprehensive Housing Service*, London: Department of the Environment

Johnson, T.J. (1972) *Professions and Power*, London: Macmillan

Kay, H. and Hardin, J. (1992) *The Rent to Mortgage Scheme in Scotland*, Edinburgh: Scottish Office

Laffin, M. (1986) 'Professionalism and Policy', *The Role of the Professions in the Central Local Government Relationship*, Aldershot: Gower

Lansley, S., Goss, S. and Wolmar, C. (1989) *Councils in Conflict: The Rise and Fall of the Municipal Left*, London: Macmillan

Le Grand, J. and Bartlett, W. (eds) (1993) *Quasi-Markets and Social Policy*, London: Macmillan

Local Government Training Board (1987) *Getting Close to the Public*, Luton: LGTB

Lowe, S. (1986) *Urban Social Movements: The City After Castells*, London: Macmillan

Lukes, S. (1974) *Power: A Radical View*, London: Macmillan

REFERENCES

Lukes, S. (ed.) (1986) *Power*, Oxford: Basil Blackwell
Lupton, T. and Mitchell, D. (1954) 'The Liverpool Estate', in Mitchell, G.D. *et al.* (eds) *Neighbourhood and Community*, Liverpool: Liverpool University Press
Maclennan, D. *et al.* (1989) *The Nature and Effectiveness of Housing Management in England*, London: HMSO
Madge, J. and Brown, C. (1981) *First Homes: A Survey of the Housing Circumstances of Young Married Couples*, London: Policy Studies Institute
Maud (1967) *Report of the Committee on the Management of Local Government*, London: HMSO
Merrett, S. (1979) *State Housing in Britain*, London: Routledge
Moorhouse, B., Witson, M. and Chamberlain, C. (1972) 'Rent strikes – direct action and the working class', in Miliband, R. and Savile, J. (eds) *The Socialist Register*, London: Merlin, pp.133–56
Morris, R. and Mogey, J. (1954) 'The Sheffield Estate', in Mitchell, G.D. *et al.* (eds) *Neighbourhood and Community*, Liverpool: Liverpool University Press
Murray, C. (1990) 'The emerging British underclass', *Choice in Welfare No. 2*, London: Institute of Economic Affairs
National Consumer Council (1986) *Measuring Up: Consumer Assessment of Local Authority Services: A Guideline Study*, London: NCC
Newton, K. (1976) *Second City Politics*, London: Oxford University Press
Niskanen, W.A. (1971) *Bureaucracy and Representative Government*, Chicago: Aldine-Atherton
Niskanen, W.A. (1973) *Bureaucracy: Servant or Master*, London: Institute of Economic Affairs
Oppenheim, C. (1990) *Poverty: The Facts*, London: Child Poverty Action Group
Parker, T. (1985) *The People of Providence: A Housing Estate and Some of its Inhabitants*, London: Hutchinson
Peters, T. and Waterman, R.H. (1982) *In Search of Excellence*, New York: Harper and Row
Pinto, R. (1993) *The Estate Action Initiative: Council Housing Renewal, Management and Effectiveness*, Aldershot: Avebury
Power, A. (1984) *Local Housing Management*, London: Department of the Environment
Power, A. (1987) *Property before People*, London: George Allen and Unwin
Raynsford, N. (1992) 'Arm's length companies: an option for local authority housing', *Housing Review* 41.2, pp.26–8
Rhodes, R. (1987) 'Developing the public service orientation', *Local Government Studies*, Vol 13, May/June, pp.63–73
Richardson, A. (1977) *Tenant Participation in Council Housing Management*, London: Department of the Environment
Richardson, A. (1983) *Participation*, London: Routledge
Ridley, N. (1988) *The Local Right – Enabling Not Providing*, London: Centre for Policy Studies
Robinson, D. (Chairman) (1977) *Remuneration of Councillors: Vol I, Report* (Cmnd 7010); *Vol II, The Surveys of Councillors and Local Authorities*, London: HMSO

Satsangi, M. and Kearns, A. (1992) 'The use and interpretation of tenant satisfaction surveys in British social housing', *Environment and Planning, C: Government and Policy* Vol 10, pp.317–31

Saunders, P. (1979) *Urban Politics: A Sociological Interpretation*, London: Hutchinson

Saunders, P. (1990) *A Nation of Home Owners*, London: Hutchinson

Scott, S. *et al.* (1994) *Training for Tenant Management*, London: HMSO

Scottish Consumer Council (1981) *Tenants Associated*, Edinburgh: SCC

Sklair, L. (1975) 'The struggle against the Housing Finance Act', in Miliband, R. and Savile, J. (eds) *The Socialist Register*, London: Merlin, pp.251–92

Social and Community Planning Research (1988) *Tenants' Attitudes to Landlord Services* (unpublished mimeo), London: SCPR

Stewart, J. (1988) *The New Management of Housing Departments*, Luton: Local Government Training Board

Stewart, J. (1989) 'Changing organisation of local authorities', in Stewart, J. and Stoker, G. (eds) *The Future of Local Government*, Basingstoke: Macmillan

Stewart, J. and Clarke, M. (1987) 'The public service orientation: issues and dilemmas', *Public Administration* 65, pp.161-77

Stoker, G. (1989) 'Local government for a post-Fordist society', in Stewart, J. and Stoker, G. (eds) *The Future of Local Government*, Basingstoke: Macmillan

Stoker, G. (1991) *The Politics of Local Government*, London: Macmillan

Tucker, J. (1966) *Honourable Estates*, London: Gollancz

Ward, C. (1974) *Tenants Take Over*, London: Architectural Press

Widdicombe, D. (Chairman) (1986) *Report of the Committee of Inquiry into the Conduct of Local Authority Business* (Cmnd 9797): *Research Vol I, The Political Organisation of Local Authorities* (Cmnd 9798); *Research Vol II, The Local Government Councillor* (Cmnd 9799), London: HMSO

Wilcox, S. (1994) *Housing Finance Review 1994/95*, York: Joseph Rowntree Foundation

Wilcox, S. with Bramley, G., Ferguson, A., Perry, J. and Woods, C. (1993) *Local Housing Companies: New Opportunities for Council Housing*, York: Joseph Rowntree Foundation

Wilding, P. (1982) *Professional Power and Social Welfare*, London: Routledge

Winkler, F. (1988) 'Consumerism in health care: beyond the supermarket model', *Policy and Practice* 15, pp.1–8

Young, K. and Davies, M. (1990) *The Politics of Local Government since Widdicombe*, York: Joseph Rowntree Foundation

INDEX

allocation policy 41, 69, 110, 170
Almond, G. 118
Andrews, L. 115, 116
asbestos 138, 140
Association of London Housing
   Estates (ALHE) 18
Association of Women House
   Property Managers 57
Audit Commission 17, 59

Bachrach, D. 158
Baratz, M. 158
Barron, J. 77, 81
Bartlett, W. 201
Bines, W. 14, 15, 18, 20, 29, 48,
   60, 62, 105, 148
Birks, D. 112
Birmingham 23
Bloch, A. 78
Brion, M. 56
Brown, C. 108
Buchanan, J. 29

Cairncross, L. 19, 20, 32, 116,
   117, 127, 148, 164, 175, 197
Castells, M. 132
CCT see compulsory competitive
   tendering
Central Government Control
   4–23, 131, 171, 199, 201
Central Housing Advisory
   Committee: (1959) 58, 59
Chartered Institute of Housing 51,
   53, 57, 59

Citizens Charter 176
citizenship local authorities
   characteristics 27, 30–2; applied
   to participation 42–6
City University 59
Clapham, D. 11, 13, 16, 19, 20,
   32, 56, 102, 108, 116, 117,
   127, 148, 164, 175, 186, 189,
   191, 194, 197, 198
Clarke, M. 30, 31
Clegg, S. 25, 157, 159, 160, 161,
   165, 166, 172, 176
Cohen, R. 195
Cole, I, 51, 61
Clinton, A. 17
Commission for Racial Equality
   (CRE) 169
Commission for Social Justice 198
Committee on Local Authority
   and Allied Personal Social
   Services (1968) 58
Community Charge 5
Community Ownership 16–17;
   Scotland 22
compulsory competitive tendering
   (CCT) 7–8, 14, 17–18, 22, 42,
   62, 176, 186, 196, 197–8,
   200–1
Conservative party 4, 9–10,
   15–16, 39, 41, 42, 46, 87–91,
   98, 171, 174–5, 188, 196
consumerist local authorities:
   characteristics 27–9; applied to
   tenant participation 40–2

Lightning Source UK Ltd.
Milton Keynes UK
UKOW05f0607181013

219270UK00001B/2/A